P9-AGA-528

NEWTON'S DREAM

PHILOSOPHIÆ NATURALIS PRINCIPIA MATHEMATICA

Autore *J S.* NEWTON, *Trin. Coll. Cantab. Soc.* Matheſeos Profeſſore *Lucaſiano*, & Societatis Regalis Sodali.

IMPRIMATUR·
S. PEPYS, *Reg. Soc.* PRÆSES.
Julii 5. 1686.

LONDINI,

Juſſu *Societatis Regiæ* ac Typis *Joſephi Streater*. Proſtat apud plures Bibliopolas. *Anno* MDCLXXXVII.

Newton's Dream

Edited by MARCIA SWEET STAYER

Consulting Editor BORIS CASTEL

Published for *Queen's Quarterly* by McGill-Queen's University Press

ISBN 0-7735-0689-6
Legal deposit 2nd quarter 1988

Canadian Cataloguing in Publication Data

Main entry under title:

Newton's dream

Based on lectures and papers given at a conference held at Queen's University, in 1987, and celebrating the tercentenary of the publication of Newton's Principia Mathematica.

ISBN 0-7735-0689-6

1. Newton, Isaac, Sir, 1642-1727 - Knowledge - Science - Congresses. 2. Newton, Isaac, Sir, 1642-1727. Principia - Congresses. 3. Science - History - Congresses. I. Sweet Stayer, Marcia

QA803.N49 1988 501 C88-090266-3

Contents

Acknowledgements

The Organizing Committee of the *Principia* Celebration Symposium extends its appreciation to the Royal Society of Canada and to the Principals of the Royal Military College of Canada and Queen's University for their support. They would like to thank Ms Marcia Sweet Stayer for editing the papers and seeing them through publication under the auspices of *Queen's Quarterly*: Professor Clive Thomson, her co-editor, and Mrs Alice Davey, Editorial Assistant, offered valuable advice on editorial matters. Professor Boris Castel supplied much appreciated counsel throughout the publishing process. Mrs Linda Freeman expertly typed the manuscript. Mr Peter Dorn very capably designed the handsome posters and programmes, and this book. Finally, the Committee recognizes the contribution of Mr M.C. Urquhart at whose suggestion the Symposium was planned.

Preface

In 1687 Isaac Newton published *Philosophia Naturalis Principia Mathematica* – *Principia* – a work which profoundly changed the perspective with which we view the world and the universe. The importance of the work was immediately recognized not only by other scientists but by his intellectual contemporaries outside the sciences, who enthusiastically incorporated his methodology into their studies.

The 550-page treatise was the culmination of a lifetime of intellectual adventure. Newton's work prior to *Principia* had been largely unpublished but richly varied. He had earlier redirected science by inventing calculus which allowed him to explain the behaviour of bodies mathematically. He conducted serious, original, and successful research into optics, tides, thin films, and gravitation. The extraordinary success of the *Principia* lent prestige to science and its method, which for the first time possessed the theoretical structure to make predictions.

During 1987 Queen's University, the Royal Military College of Canada, and the Royal Society of Canada sponsored a series of lectures in Kingston, Ontario, as part of the Canadian Celebration of the Tercentenary of Newton's *Principia*. The speakers were distinguished visiting scholars in science, the social sciences, and the humanities, and their papers examined the impact of Isaac Newton's thought from a variety of perspectives. They met with students and other scholars in a two-day scientific symposium and delivered a series of public lectures over the course of the year. The public lectures were so well received by the educated non-specialist audiences that the organizing committee decided to make them available to more readers.

Newton's influence has barely waned and inspires contemporary scientists to try to fulfill his dream of uncovering the governing principles for all constituents of the physical world. The essays which follow eloquently convey the excitement that was felt 300 years ago at the publication of the *Principia*, and the ongoing importance of Newton's thought.

Newton and the Scientific Revolution

RICHARD S. WESTFALL

Isaac Newton published *Philosophiae naturalis principia mathematica: The Mathematical Principles of Natural Philosophy* in July 1687. Seldom has the significance of a book been more immediately recognized. Indeed, its recognition began even before publication. In the spring of 1687, Fatio de Duillier, a young Swiss mathematician who would play a central role in Newton's life during the following six years, arrived in London. He found the learned community aflutter in expectation of the book which was destined, they told him, to remodel natural philosophy (Fatio 167-69). Similarly the *Philosophical Transactions* of the Royal Society carried a review of the *Principia*, which was in keeping with Fatio's report, shortly before the publication of the book itself. Although it was not signed, the review was composed by Edmund Halley, who knew what he was talking about since he was in fact the publisher. "This incomparable Author having at length been prevailed upon to appear in public," Halley began the review, "has in this Treatise given a most notable instance of the extent of the powers of the Mind; and has at once shown what are the Principles of Natural Philosophy, and so far derived from them their consequences, that he seems to have exhausted his Argument and left little to be done by those that shall succeed him." After summarizing the contents of the book, Halley concluded in the same vein: "it may be justly said, that so many and so valuable Philosophical Truths as are herein discovered and put past Dispute were never yet owing to the capacity and industry of any one man" (review of *Principia*).

It was not long after publication when John Locke, who was then resident in the Netherlands as a political refugee from the regime of James II, heard about the work. Unable to cope with its difficult mathematics, Locke asked Christiaan Huygens whether he could trust the book, and with Huygens's assurance that he could, Locke applied himself to the prose.[1] When he returned to England in the wake of the Glorious Revolution, Locke made it one of his first items of business to form Newton's acquaintance. The learned world in England did not lag behind Locke in acknowledging the *Principia*, so that its author vaulted in one leap from relative obscurity to the position of leadership among English thinkers.

On the Continent the overt indicators of the *Principia*'s reception differed, because continental philosophers had fundamental objections to its concept of attractions. For all that, they did not fail to recognize the power of the book, and they found themselves wholly unable to ignore it. It is indicative of the *Principia*'s impact on the Continent that when the French reorganized the Académie des Sciences in 1698, they made Newton one of the eight original foreign associates (Cohen, "Isaac Newton"). In a word, there has never been a time when the *Principia* was not seen as an epochal work, and there has never been a time since its publication when Newton was not perceived as one of humanity's leading intellects, much more than merely a genius. Although none of the statements I have quoted explicitly say as much, it also seems correct to me to say that there has never been a time when Newton's greatness was not seen to be associated with the fact that he did not stand alone, that he came after Copernicus, Kepler, Galileo, Descartes, Huygens, and numerous others. That is, Newton has always been recognized as the climax of the intellectual movement we call the scientific revolution of the seventeenth century, and that recognition defines the task I have set myself in this paper – to give an account first of the scientific revolution and then of the relation of Newton and his *Principia* to it.

Beyond the ranks of historians of science, in my opinion, the scientific revolution is frequently misunderstood. A vulgarized conception of the scientific method, which one finds in elementary textbooks, a conception which places overwhelming emphasis on the collection of empirical information from which theories presumably emerge spontaneously, has contributed to the misunderstanding, and so has a mistaken notion of the Middle Ages as a period so absorbed in the pursuit of salvation as to have been unable to observe nature. In fact, medieval philosophy asserted that observation is the foundation of all knowledge, and medieval science (which certainly did exist) was a sophisticated systematization of common sense and of the basic observations of the senses. Modern science was born in the sixteenth and seventeenth centuries in the denial of both.

Consider astronomy for example. Nearly everyone takes Copernicus as the beginning of the scientific revolution, and developments in astronomy foreshadowed the course of the scientific revolution as a whole.[2] Medieval astronomy rested on two basic propositions: that the motions observed in the heavens actually take place there, and that we must accept the validity of the basic observation each of us makes every moment, to wit, that we live on a stable earth. Geocentric astronomy followed directly from these premises, a complex system of circles on circles, of deferents, epicycles, and eccentrics, which provided a reasonable account in theory of the observed celestial phenomena.

Early in the sixteenth century, Nicholas Copernicus became dissatisfied with the system. It contained arbitrary elements. For example, in geocentric astronomy the sun was only one planet among seven, and yet the sun was involved in the theories of all the others except the moon. Mercury and Venus never depart far from the sun; they are seen in the west in the evening after sunset, or in the east in the morning before dawn, but never in the midnight sky. In order to make the theories of Mercury and Venus work, the centres of their epicycles had always to lie on the line between the sun and the Earth. Mars, Jupiter, and Saturn, on the other hand, go through their retrogressions when they are in the opposite part of the heavens from the sun; each of them, in the middle of its retrogression, crosses the meridian precisely at midnight. To make their theories correspond to the heavens it was necessary that the *radii vectores* of their epicycles always be parallel to the line between the sun and the Earth. Why was this so? How was it possible that the sun be merely one planet and yet participate in the theories of the others? Moreover, the geocentric system contained no necessary criteria for orbital size; as long as deferent and epicycle maintained the necessary proportion to each other established by observation, they could be of any size. Hence the planets had no necessary order. The Ptolemaic order was commonly accepted, but only on the basis of *ad hoc* assumptions that had no intrinsic relation to the system; there was no conclusive argument why Saturn, for example, could not be closer to the Earth than Mercury. And the so-called system did not appear to be a system at all to Copernicus. Consider the order of motions, in the Ptolemaic arrangement, as Copernicus would have known them. The Earth stood motionless in the centre. The moon circled the Earth with a period of one month. Beyond the moon were Mercury, Venus, and the sun, all with periods (for Mercury and Venus, average periods), in the Ptolemaic system, of one year. Then Mars, two years, Jupiter, twelve years, Saturn, thirty years, and beyond all the planets the sphere of the fixed stars with a period of one day in the opposite direction. In Copernicus's eyes, this was not an ordered system, it was chaos.

Copernicus saw that he could remove all of the arbitrary elements and solve all of the problems by the simple expedients of putting the Earth into motion, with both a diurnal rotation and an annual revolution around the sun, and of treating most of the motions in the heavens as mere appearances resulting from the motions of the Earth. All that was necessary was to put the Earth in motion; this was an incredible thought, fundamentally at odds with all experience and with the dictates of common sense. And yet, by accepting that premise, he could arrive at a system that presented a spectacle of mathematical order and harmony not to be found in Ptolemy. For Ptolemaic, geocentric, astronomers, each planet was a separate problem. For Copernicus and heliocentric astronomers, system was foremost. In the heliocentric system, the orbits are measurable in terms of the astronomical unit (the distance, not well measured in terrestrial units in Copernicus's age, between the Earth and the sun). Hence they had a necessary order, and that order corresponded to a harmonious system of motions in which periods decreased with distance from the sun, and the fixed stars at the periphery stood motionless. Copernicus threw common sense to the winds in order to pursue the arcane satisfactions of mathematical harmony (Copernicus, *Revolutions*).[3]

Copernicus was only the beginning. If he challenged some of the assumptions of common sense, he was unable to recognize the others to which he still clung. Foremost among these was the conviction that only circular motions can be found in the heavens and thus that astronomy can employ only circles in its account of heavenly phenomena. To the ancient Greeks, the circle had represented the perfect figure, that path in which a body can move forever without altering its relation to the centre, and the perfection of the circle had seemed to correspond to the perfection and immutability of the heavens. Copernicus was as wedded to the notion that the circle was the sole device of astronomical theory as the ancient Greeks, but circles obstructed the pursuit of a mathematically simple and harmonious system. It remained for Johannes Kepler, two generations later, to challenge their role. As Kepler began to propose that the planets move in non-circular orbits, he received a letter of protest from another astronomer who insisted that he was destroying the very foundation of astronomy. In a witty reply, Kepler referred to circles as voluptuous whores enticing astronomers away from the honest maiden Nature (Kepler, "Letter" 205). He knew whereof he spoke, for it had taken Kepler years to escape the attractions of the enchantress. His elliptical orbits, together with his two other quantitative laws of planetary motion, which yielded a system breathtaking in its mathematical simplicity, was another victory of abstract reason over assumptions accepted for centuries as the very embodiment of common sense.[4]

Kepler's three laws did nothing, however, to make the proposition that the Earth is in motion one whit less incredible than it seemed to virtually everyone. Here we confront the phenomena of moving objects, especially falling objects, on an Earth said to be rotating on its axis. The objections raised were by no means silly, and once again they sprang from common sense itself and from the observations all of us constantly make. If the Earth is moving as Copernicans claimed, surely we would perceive the motion. The size of the Earth was known then with sufficient accuracy to make the point. If it is rotating on its axis, we are, at this moment, moving from west to east at approximately a thousand miles per hour. To put the dilemma in twentieth-century terms, which make the problem considerably less difficult than seventeenth-century conditions, we ride in our cars every day, and we never fail to perceive the motion. Is it possible that we are cruising down the cosmic highway at a rate well over ten times the highest speed we ever go in our cars and are yet unaware of the motion? Or drop a stone from the top of a tall building with a flat side. Ballantine Hall on the campus of Indiana University – the twentieth-century example in this case differs from a seventeenth-century one only in the style of the architecture – is nine storeys tall; it takes a stone roughly two and a half seconds to fall from the top to the ground. A thousand miles per hour is equivalent to more than a thousand feet per second. During the time the stone is falling, Ballantine Hall moves more than half a mile to the east. How is it possible that a stone dropped one foot out from the eastern wall of Ballantine falls parallel to its side and lands very nearly one foot from the base of the wall?

The problem of motion on a moving Earth defined the major work of Galileo, who rethought the very conception of motion in order to justify the assertion that the Earth is turning on its axis. Motion, Galileo decided, is not, as Aristotle had thought, a process whereby entities realize their being. Motion is merely a state in which a body finds itself, a state that alters nothing in a body, a state to which a body is indifferent. Hence we are unable to perceive uniform motions in which we participate along with everything around us. When motion is understood in these terms, stones can fall parallel to the vertical walls of buildings when the Earth is in motion as well as they would if the Earth were at rest. Contrary to the central assertion of Aristotle's analysis, uniform motion understood in Galileo's terms requires no cause. As I hold the stone before I drop it from the top of Ballantine Hall, the stone is moving from west to east at the same rate as the building and I. The stone's horizontal motion continues unaffected; the eastern wall of Ballantine does not catch up with the stone. We perceive only the vertical drop in which we do not participate. As Descartes, who shared the new conception, put it, philosophers have been asking the wrong question. They have been asking what keeps a body in motion. The correct question is why does it not continue to move forever (*Oeuvres*). We know the new conception of motion as the principle of inertia. Although Galileo did not use that term, he made the concept the corner-stone of a new science of motion (or mechanics, as physicists call it) which became the central edifice in the whole new complex of modern science, and philosophers of science today agree that the principle of inertia is the basic concept on which the science we know rests.[5]

With Galileo the Copernican universe became believable, but was it true? Or to rephrase the question, what evidence was there in its favour in the early decades of the seventeenth century? As soon as one puts the question in those terms, one is forced to concede that the evidence in its favour was almost precisely the advantage that Copernicus, and Copernicans after him, had pursued, that is, mathematical harmony and simplicity. For the truth is that there was precious little other evidence to support it. Galileo's new science of motion answered the major objection against the system, but it did not count as evidence for it. To be sure, late in 1609 the same Galileo had turned his newly improved telescope on the heavens, but neither Galileo nor anyone else could look through a telescope and see the Earth moving around the sun and rotating on its axis. Galileo did observe the mountainous surface of the moon, the spots on the sun from the motion of which he inferred the rotation of the sun on its axis, and the satellites of Jupiter. These phenomena all fit more smoothly into the Copernican picture of the universe, but they did not demonstrate that it was true. Especially Galileo observed the phases of Venus which did demonstrate that Venus (and by implication Mercury) revolves around the sun and were then logically incompatible with the Ptolemaic system, but the phases of Venus did not demonstrate that the Copernican system is true.[6] One forgets all too easily another observation that Galileo did not make, stellar parallax. If the Earth is moving around the sun in an immense orbit, then if we observe the angle at which some fixed star appears in the middle of the summer and again six months later in the middle of the winter, after the Earth has moved an immense distance, surely the two angular locations will differ. To the naked eye the angles appeared identical. Alas, they also appeared identical through early telescopes. Today we know why: the fixed stars are so far removed that it was well into the nineteenth century before telescopes powerful enough to distinguish the two angles were developed. Such distances were inconceivable to most people in the early seventeenth century. At the least, the failure to observe stellar parallax offset the positive observation of the phases of Venus. Primarily for the ethereal advantages of mathematical harmony and simplicity early scientists asked mankind to surrender the most obvious evidence of the senses and the manifest dictates of common sense. They did not ask in vain. During the seventeenth century a new school of natural philosophers who were ready and indeed eager to accept the invitation appeared until, by the end of the century, none of them could imagine how anyone had ever believed otherwise.

Observe the process that I describe. From Copernicus's question about the order of the universe there spread out an expanding domain of discussion that grew ever broader. It does not appear to me to have been spurred on by issues of practical utility as so much scientific investigation in our age is, but rather by the pursuit of Truth. (I capitalize "truth" as I am convinced sixteenth- and seventeenth-century natural philosophers would have done.) Johannes Kepler, a man without personal resources, who was dependent for his livelihood on the favours of patrons and, conscious as he was of the current patron's mortality, always on the look-out for the next, was willing to live teetering on the brink of oblivion if he could demonstrate the correct pattern of the heavens. Galileo was willing to dare the fury of the Inquisition because it mattered to him whether the universe was geocentric or heliocentric.

The next major step was taken by a Frenchman, René Descartes, to whom such issues also mattered. Instructed in part by Galileo's debacle, he chose to live in exile in the Netherlands in order that he might more freely pursue his thoughts. Descartes universalized the tendencies inherent in the scientific movement. It is not only the heavens that are not as they seem to be, and not only motion. The whole universe is not as it seems to be. We see about us a world of qualities and of life. They are all mere appearances. Reality consists solely of particles of matter in motion. Some of the particles impinge on our senses and produce sensations, but nothing similar to the sensations exists outside ourselves. Reality is quantitative, particles characterized solely by size and shape, and of course motion. The picture of nature that I am describing so briefly was known in the seventeenth century as the mechanical philosophy; it provided the philosophic framework of the scientific revolution. The world was not like a living being, it was like a great machine. Things such as plants and animals, said to be living – human beings, with their capacity for rational thought were partial exceptions – were only complicated machines. Thus the ultimate implication of the movement Copernicus initiated proved to be much wider than Copernicus had imagined. Humankind was not merely displaced from the centre. It was displaced entirely. Its presence was irrelevant to the universe; the universe was not created for human benefit: it would have been almost entirely the same whether humankind was there or not.[7]

Isaac Newton arrived on the scene after the early stages of the scientific revolution. He was born in 1642, the year Galileo died, twelve years after Kepler died, two years before Descartes published his *Principles of Philosophy*. Everything that I have discussed contributed to his intellectual inheritance. The scientific revolution was a complex movement of many dimensions. For my purposes in this paper, let me summarize it under three major themes, all of crucial importance to Newton. The scientific revolution presented a new picture of the universe, heliocentric rather than geocentric. It presented a new image of nature as inert particles in motion, the mechanical philosophy. And it presented a new vision of reality in terms of quantity rather than quality, so that increasingly science would express itself through mathematical demonstrations. Though implicit in the pursuit of mathematical harmony and simplicity, the third theme has not been explicit in the examples I have cited. To insist on its centrality, let me quote two representative statements. "Geometry," said Kepler, "being part of the divine mind from time immemorial, from before the origin of things, being God Himself (for what is in God that is not God Himself?), has supplied God with the models for the creation of the world and has been transferred to man together with the image of God. Geometry was not received inside through the eyes."[8] The same vision was shared by Galileo, who brought geometry down from the heavens, and building with it on the foundation of his new conception of motion, created what had heretofore appeared to be a contradiction in terms, a mathematical science of terrestrial motion. "Philosophy," said Galileo, "is written in this grand book, the universe, which stands continually open to our gaze. But the book cannot be understood unless one first learns to comprehend the language and read the letters in which it is composed. It is written in the language of mathematics, and its characters are triangles, circles, and other geometric figures without which it is humanly impossible to understand a single word of it ... " (*Discoveries* 237-8). The new vision of reality, the conviction that the world is structured mathematically, was perhaps more basic to the scientific revolution than anything else.

When Newton enrolled in Cambridge in 1661, none of the early work of the scientific revolution had penetrated the standard curriculum. Despite some superficial changes, the university remained still in its medieval mould, and the philosophy of Aristotle continued to be the focus of the studies it prescribed. Newton was not a docile student, however, and sometime around 1664 he undertook a new programme of study. The record of his reading survives.[9] It contains no suggestion of tutorial guidance; on the contrary, it strongly implies that Newton struck out on his own. First, apparently, he discovered mathematics. In this he may have been influenced by the inaugural lectures of Isaac Barrow in the newly created Lucasian Professorship of Mathematics. Barrow was not Newton's tutor, however, and the two men became close only five years later. The seventeenth century was the most creative period in mathematics since the age of ancient Greece; in the short space of twelve to fifteen months, during what was also his final year as an undergraduate, Newton, by himself, absorbed the entire prior achievement of seventeenth-century mathematics and began to move beyond it towards the calculus. He set down the definitive statement of what he called the fluxional method in October 1666, about one and a half years after he had taken his Bachelor of Arts degree.[10] Mathematics was not all he discovered. He also found the new natural philosophy, especially the writings of Descartes and Pierre Gassendi (whose revival of ancient atomism offered an alternative mechanical philosophy). Within the context of natural philosophy, he came upon the problem of colours. When he was less than a year beyond his BA degree, Newton set down in the same notebook in which he recorded his reading in natural philosophy as a whole the first suggestion of the central concept to which all of his work in optics would be devoted, that light is not homogeneous as everyone had heretofore believed, but a heterogeneous mixture of difform rays that provoke different sensations of colour.[11] During much the same period, Newton also discovered the science of mechanics.[12] One of the best known Newtonian stories, which is drawn from a passage he composed toward the end of his life, claims that he developed the concept of universal gravitation at this time. Newtonian scholars no longer believe that story. Nevertheless, it is clear that he entertained vague thoughts about the dynamics of the heavens at this time, and that he later drew upon and amplified these thoughts (Herivel 183-98).

Later Newton found other interests as well, primarily chemistry/alchemy, and theology, which together largely dominated his time and consciousness from the late 1660s. In August 1684, as even people only vaguely informed about Newton know, he received an unexpected visit from Edmund Halley, who bluntly asked him what would be the shape of the path followed by a body orbiting another that attracted it with a force that varied inversely as the square of the distance. We cannot follow in any depth the psychology of Newton's response to Halley's question. We can only say that somehow the question reawakened earlier interests long dormant, so that late in the same year Newton sent Halley a tract of ten pages, which is known by the title *De motu*.

De motu was a short treatise on orbital dynamics that demonstrated the relation of Kepler's three laws of planetary motion to an inverse square attraction (Herivel 257-92). Although the demonstrations in *De motu* later became part of the *Principia*, they did not rest on a solid foundation in the early tract, for it presented only a primitive and crude science of dynamics. It did not accept the principle of inertia, and it did not state any general force law. Moreover, the defects of *De motu* were not so much the shortcomings of Newton as the limitations of the science of his day. In 1684, no one had constructed a satisfactory science of dynamics that corresponded to the new conception of motion. In Kepler's conclusions about planetary motion and in Galileo's science of uniformly accelerated motion, mechanics had arrived at kinematic laws, but as yet no dynamics to support the kinematics existed. Newton's first task, as he set out to expand and elaborate *De motu*, was to create a workable science of dynamics.

Those papers also survive (Herivel 292-320). They show that during the months following the composition of *De motu* Newton carried out an intense investigation of the fundamental concepts of the science of mechanics. Partly this was a work of definition, and as he proceeded, Newton provisionally defined no less than nineteen different concepts, of which five definitions survived into the *Principia*. From the investigation emerged the three laws of motion that still introduce courses in physics just as they introduced the *Principia* three centuries ago, three laws that constitute a quantitative science of dynamics from which both Galileo's kinematics of uniformly accelerated motion and Kepler's kinematics of planetary motion follow as necessary consequences. Thus Newton's science of dynamics bound celestial mechanics together with terrestrial mechanics, and for this reason people frequently refer to it as the Newtonian synthesis.

As an expression of Newtonian dynamics, the *Principia* was also a synthesis of the major themes of the scientific revolution. It supplied the final justification of the heliocentric system of the universe, in its Keplerian form, by providing its dynamic foundation. It raised the mathematical vision of reality to a new level of intensity. The *Principia* was a book of mathematical science, modelled on Euclid; more than any other work of the scientific revolution, it established the pattern that the whole of modern science has striven to fill in. The third theme that I singled out, the mechanical image of nature, is perhaps less obvious in the *Principia*; nevertheless, the book is unintelligible apart from it. Like every mechanical philosopher in the seventeenth century, Newton looked upon nature as a system of material particles in motion. To the astringent ontology of the prevailing mechanical philosophy, which ascribed to particles only size, shape, and motion, and insisted that all the phenomena of nature are produced by impact alone, Newton added a further category of property – forces of attraction and repulsion – whereby particles and bodies composed of particles act upon other particles and bodies at a distance.[13] The Newtonian conception of attractions also formed an essential dimension of Newton's mathematical view of reality, for the forces were mathematically defined. For example, the gravitational attraction of bodies for each other varied inversely as the square of the distance.

The structure of the *Principia* reflects its dynamic foundation. Book I is all abstract mathematical dynamics of point masses moving without resistance in various force fields. While Newton consistently explored the consequences of different force laws in the problems to which Book I devoted itself, he focussed his attention primarily upon the inverse square attraction and the phenomena of motion it entails. Book II considered the motion of bodies through resisting media and the motions of such media, reaching its climax with the examination of the dynamic conditions of vortical motion. That is, Book II was primarily an attack on the natural philosophy of Descartes and the prevailing mechanical picture of nature he had inspired. Newton demonstrated that vortices are unable to sustain themselves without the constant addition of new energy (or "motion," in the language of the seventeenth century) and are unable to yield Kepler's three planetary laws.

With the alternative system discredited, Book III returned to the demonstrations of Book I, which it applied to the observed phenomena of the universe. Newton's first law states that bodies in motion, undisturbed by any external influence, tend to move in straight lines. Planets move in closed orbits, and in their motions they observe Kepler's second law, the law of areas. For both reasons it follows from Book I that the planets must be attracted toward a point near the centre of their orbits where the sun is located. For the same reasons, the satellites of Jupiter and Saturn must be attracted toward those planets. Moreover, planets travel in ellipses with the sun at one focus, and their lines of apsides (the major axes of the ellipses) remain stable in space. The orbits of the system also obey Kepler's third law. It follows from the demonstrations of Book I that the attraction toward the sun must vary inversely as the square of the distance, and since the satellites of Jupiter conform to Kepler's third law, they too must be attracted by such a force. Furthermore, because the planetary system observes Kepler's third law, the attraction of the sun for each of the planets must be proportional to its quantity of matter. Again, the same holds for the attraction of Jupiter for its satellites, and since the satellites move in orbits that are nearly concentric with Jupiter as it orbits the sun, the attraction of the sun for both Jupiter and its satellites must be proportional to their several quantities of matter.

The Earth also has a satellite, the moon, which constantly accompanies it and does not fly off into space along a straight line; therefore the Earth must attract the moon to hold it in its orbit. Unlike Jupiter, the Earth has only one satellite, and that one with an orbit so highly irregular as not to appear truly elliptical; from such a satellite alone one cannot reason to an inverse square force. However, on the Earth there is a substitute for additional satellites. Heavy bodies fall to its surface where we can experiment with them. Hence the importance of the correlation between the motion of the moon and the measured acceleration of gravity on the surface of the Earth. How far from the Earth is a heavy body we drop – or, to speak in terms of the famous myth, how far is an apple from the Earth? Whether it was an apple or an experimental weight, any body that a seventeenth-century scientist could handle was at most a few feet from the surface of the Earth. Newton's correlation demanded that it must be roughly four thousand miles from the Earth; that is, the crucial distance was not to the surface but to the centre. To put it this way is to insist on the importance of the demonstration at which Newton arrived some time in 1685, that a homogeneous sphere, composed of particles that attract inversely as the square of the distance, itself attracts other bodies, no matter how close they may be, with forces that are inversely proportional to the bodies' distances from the sphere's centre. With the crucial demonstration about spheres, Newton could further demonstrate that the attraction that causes heavy bodies to fall toward the Earth is quantitatively identical to the attraction that holds the moon in its orbit. The critical correlation between the motion of the moon and the acceleration of gravity was the single strand that connected the cosmic attraction, shown to be necessary to hold the solar system together, with terrestrial phenomena, thus allowing him to apply to the attraction the ancient word *gravitas*, heaviness. From the motions of pendulums he could show that heaviness on Earth also varies directly in proportion to the quantity of matter. Thus Newton was brought to state what is perhaps the most famous generalization of modern science, that "there is a power of gravity pertaining to all bodies, proportional to the several quantities of matter which they contain" (*Mathematical Principles* 414).

The principle of universal gravitation, which I have just quoted, appeared early in Book III as Proposition 7, derived from the sharply limited number of phenomena I cited. The rest of the book then applied the principle to a number of other phenomena that had not contributed to its derivation. Recent European expeditions, especially a French expedition to the northern coast of South America, had revealed that the length of a pendulum that completes a swing in one second varies with the latitude. Near the equator a seconds pendulum needs to be shorter than it is in Europe. Newton was able to demonstrate that the shorter length of the pendulum is due to the decrease in the intensity of gravity near the equator because of the oblate shape of the Earth.

He turned from the shape of the Earth to the perturbations of the moon. Over the centuries astronomers had empirically established a number of anomalies in the moon's motion. Newton now showed that all the known perturbations are dynamic effects of the attraction of a third body, the sun. He applied the same analysis of the effects of a third body to the shape of a ring of water treated as a satellite circling the Earth and being perturbed by the combined effects of the moon and the sun, and he arrived at the explanation of the tides. When the "satellite" was the bulge of matter around the equator, the same analysis yielded the conical motion of the Earth's axis known as the precession of the equinoxes. In a final *tour de force*, Newton reversed a tradition as old as astronomy itself and, treating comets as planet-like bodies subject to the same orbital dynamics as planets, he succeeded in describing observed locations of the great comet of 1680-81 in terms of a parabolic orbit.

It would be difficult to overestimate the impact of the *Principia*. As I insisted, there has never been a time since the day of its publication when it was not perceived as a monumental achievement. Take its treatment of the moon as an example. The *Principia* suggested for the first time the cause of the moon's known anomalies, and in so doing inaugurated a wholly new chapter in lunar theory. Much the same can be said of the tides, the precession of the equinoxes, comets, and, of course, the book's central problem, planetary motion. Moreover, all of these phenomena were reduced to a single causal principle, and all was done with a degree of mathematical precision that made it impossible for anyone who understood the mathematics to doubt the theory. In the words of David Gregory, who had just finished reading the *Principia* in the late summer of 1687, Newton taught the world "that which I never expected any man should have known" (*Correspondence* 2: 484). In the more recent language of Thomas Kuhn, the *Principia* established the paradigm which modern science in its various dimensions has been attempting to emulate ever since.

The *Principia* was published three hundred years ago. In comparison to the total population of Europe, it was, to be sure, only a handful who recognized its significance. Nevertheless, its appearance was the most important event of 1687. I am not exactly an impartial judge, but let me state forthrightly that I am not aware of any event to match its impact on western civilization during the intervening three centuries.

NOTES

1 J.T. Desaguliers stated that he had heard this story from Newton. John Conduitt was also familiar with the story, and we can assume with confidence that he got it from the same source (King's College, Keynes mss. *130.6*, Book 2: *130.5*, Sheet 1). Newton undoubtedly learned it from Locke himself.

2 For more detailed treatment of the astronomical revolution see Koyré, *Astronomical Revolution* and Kuhn.

3 For short statements of the system see Copernicus's *Commentariolus* or Rheticus's *Narratio prima* in Copernicus, *Three Treatises*.

4 See especially Kepler's *Astronomia nova*, the work of 1609 in which he announced his first two laws, and his summary statement of his work in *The Epitome of Copernican Astronomy* in the *Gesammelte Werke*. Books IV and V of the *Epitome* have been translated into English by C.G. Wallis in vol. 16 of *Great Books*.

5 For Galileo's discussion of the problem of motion on a moving Earth see primarily the Second Day of his *Dialogue*. For his science of mechanics as a whole, see his *Two New Sciences*. There is an enormous literature on Galileo and his new science of motion. The basic work that shaped our present understanding of Galileo is Koyré, *Galileo Studies*. See also Clavelin.

6 Galileo's accounts of his observations appear in *The Starry Messenger* (1610), the opening passage of the *Discourse on Bodies in Water* (1612), and the *Letters on Sunspots* (1613). Translations of the first and the third can be found in *Discoveries*. Drake has also edited a seventeenth-century translation of the second, published as a separate volume, and more recently included his own translation of it in Galileo, *Cause*.

7 See especially Descartes's *Principles* and also the shorter exposition of his natural philosophy in *Le monde*. For discussions of the mechanical philosophy see Hall, Harré, and the relevant chapter of Collingwood.

8 *Harmonices mundi*, IV, in *Gesammelte Werke*, 6: 223. I quote the translation in Caspar, 271.

9 See his undergraduate notebook, although it contains the record only of his philosophical studies; the records of his work in mathematics and mechanics are in McGuire and Tamny.

10 The record of Newton's early studies in mathematics, up to the tract of October 1666, makes up the first volume of *Mathematical Papers*. In the notes to this edition and in other writings, D.T. Whiteside is also the leading commentator on Newton's mathematics.

11 For Newton's early work in optics see Shapiro and my own article. Shapiro is currently editing a full publication of Newton's optical papers. Volume one, which has appeared, contains the early Lucasion lectures on optics. Volume two will contain the material I am now discussing.

12 His early notes on mechanics, mostly in another notebook that Newton called the "Waste Book," are published and discussed in Herivel, 121-82.

13 Scholars argue about whether Newton meant to ascribe forces to bodies or understood that they were caused by some "mechanical" device (in the seventeenth-century meaning of "mechanical") such as a particulate aether. For the view that Newton understood forces as real entities in nature, see McGuire, "Forces," 154-208. For the argument that he understood gravity, for example, to be caused by a material medium, see Cohen, *Newtonian Revolution*. Home, 95-117, vigorously defends the position that Newton never understood electrical and magnetic forces as actions at a distance. I have generally taken the side that Newton did accept forces as real entities in nature, but the argument of this paper does not depend on that position. Whatever Newton thought about the ontological status of forces, the *Principia* proceeded in terms of attractions and repulsions acting at a distance.

WORKS CITED

Caspar, Max. *Kepler*. Trans. C. Doris Hellman. New York: Abelard-Schuman, 1959.

Clavelin, Maurice. *The Natural Philosophy of Galileo*. Trans. A.J. Pomerans. Cambridge, MA: MIT Press, 1974.

Collingwood, R.G. *The Idea of Nature*. Oxford: Clarendon, 1945.

Cohen, I.Bernard. "Isaac Newton, Hans Sloane and the Académie des Sciences." *Mélanges Alexandres Koyré*. 2 vols. Paris: Hermann, 1964. 2:61-116.

______. *The Newtonian Revolution*. Cambridge and New York: Cambridge UP, 1980.

Copernicus, Nicholas. *On the Revolutions*. Trans. Edward Rosen. Baltimore: Johns Hopkins UP, 1978.

______. *Three Copernican Treatises*. Trans. Edward Rosen. New York: Dover Publications, 1939.

Desaguliers, J.T. Preface *A Course of Experimental Philosophy*. 2 vols. London, 1734-44.

Descartes, René. *Le monde, ou traité de la lumière*. Trans. Michael Mahoney. New York: Abarus Books, 1979.

______. *Oeuvres de Descartes*. Ed. Charles Adam and Paul Tannery. 12 vols. Paris: L. Cerf, 1897-1910.

______. *Principles of Philosophy*. Trans. V.R. and R.P. Miller. Dordrecht: Kluwer, 1983.

Fatio de Duillier, Nicholas. Letter to Christiaan Huygens, 14 June *1687*. Christiaan Huygens. *Oeuvres complètes*. 22 vols. The Hague, 1888-1950. Vol. 9.

Galileo. *Cause, Experiment, and Science*. Trans. Stillman Drake. Chicago: U of Chicago P, 1981.
______. *Dialogue Concerning the Two Chief World Systems – Ptolemaic and Copernican*. Trans. Stillman Drake. Berkeley: U of California P, 1962.
______. *Discoveries and Opinions of Galileo*. Trans. Stillman Drake. Garden City, NY: Doubleday, 1957.
______. *Two New Sciences*. Trans. Stillman Drake. Madison: U of Wisconsin P, 1974.
Great Books of the Western World. Ed. Robert M. Hutchins. 50 vols. Chicago: W. Benton, 1952.
Gregory, David. Letter to Newton. 2 September 1687. Newton, *Correspondence*.
Hall, Marie Boas. "The Establishment of the Mechanical Philosophy." *Osiris* 10 (1952): 412-541.
Harré, Rom. *Matter and Method*. London: Macmillan; New York: St Martin's, 1964.
Herivel, John. *The Background to Newton's* Principia. Oxford: Clarendon Press, 1965.
Home, R.W. "Force, Electricity, and the Powers of Living Matter in Newton's Mature Philosophy of Nature." *Religion, Science, and Worldview*. Ed. Margaret J. Osler and Paul L. Farber. Cambridge and New York: Cambridge UP, 1985.
Kepler, Johannes. *Gesammelte Werke*. Ed. Walther von Dyck and Max Caspar. 19 vols. Munich, 1937-83.
______. Letter to Fabricius. 10 November 1608. *Kepler. Gesammelte Werke*. Vol. 16.
Koyré, Alexandre. *The Astronomical Revolution*. Trans. R.E.W. Maddison. Ithaca: Cornell UP; Paris: Hermann, 1973.
______. *Galileo Studies*. Trans. John Mepham. Hassocks [Engl.]: Harvester Press, 1978.
Kuhn, Thomas. *The Copernican Revolution*. Cambridge, MA: Harvard UP, 1957.
McGuire, J.E. "Forces, Active Principles, and Newton's Invisible Realm." *Ambix* 15 (1968): 154-208.
McGuire, J.E., and Martin Tamny. *Certain Philosophical Questions: Newton's Trinity Notebook*. Cambridge and New York: Cambridge UP, 1983.
Newton, Isaac. *The Correspondence of Isaac Newton*. Ed. H.W. Turnbull. 7 vols. Cambridge UP, 1959-77.
______. *The Mathematical Papers of Isaac Newton*. 8 vols. Ed. D.T. Whiteside. Cambridge and New York: Cambridge UP, 1967-81.
______. *The Optical Papers of Isaac Newton*. Ed. Alan E. Shapiro. 3 (projected) vols. Cambridge and New York: Cambridge UP, 1984.
______. *[Principia.] Sir Isaac Newton's Mathematical Principles of Natural Philosophy and his System of the World*. Ed. Florian Cajori. Trans. Andrew Motte. Berkeley: U of California P, 1934.
Rev. of *Principia*. *Philosophical Transactions of The Royal Society of London* 16 (1686-87): 291-97.
Shapiro, Alan E. "The Evolving Structure of Newton's Theory of White Light and Color." *Isis* 71 (1980): 211-35.
Westfall, Richard S. "The Development of Newton's Theory of Colors." *Isis* 53 (1962): 339-58.

Science, Rationality, and Newton

W. H. NEWTON-SMITH

In 1739 there appeared an English translation of the charming Italian work *Sir Isaac Newton's Philosophy Explained for the Use of the Ladies* by one Francesco Algarotti. It is in the form of a dialogue between the author and a discreetly unnamed marchioness. By page 170 of the second of these two volumes, the marchioness has been guided through Newton's universal law of gravitation. She herself has been led to calculate that if the distance between two bodies is increased by a factor of five, the force of gravitational attraction is diminished by a factor of twenty-five.

> I cannot help thinking, said the Marchioness, that this Proportion in the squares of the Distances of Places, or rather of Times, is observed even in Love. Thus after eight Days Absence Love becomes 54 times less than it was the first Day and according to Proportion it must soon be entirely obliterated; I fancy there will be found, especially in the present Age very few Experiences to the contrary. (170)

In fairness to our marchioness, who has previously had a lesson on the squaring of numbers, we should presume "54" to be a printing error for "64." In response, Segnior Algarotti ventured the opinion that this theorem applies to both sexes and may in fact be an inverse cube law. Smoothie that he was, he went on to suggest that our marchioness was herself so attractive that she had the power to reverse the theorem and increase the passion through time. But the marchioness was a true Newtonian. "No! no! said the Marchioness, Gallantry must never destroy a theorem. If Geometry was permitted to get some Footing here in little Time it would produce Wonders"(171).

It is quite remarkable that within a few years of Newton's death Italian marchionesses are being represented as extending the research programme of his *Principia* to love, an extension which the virginal Newton himself was singularly ill-equipped to make. Whether or not the theories of Newton as revealed in

A version of this paper has appeared in a Chinese translation in Journal of Dialectics of Nature, *9 (1987): 1-11.*

his *Principia* have the power to generate the laws of passion, there is no doubt that *Principia* remains the most powerful single scientific work yet produced. Taking that as established I look in this paper at the role of Newton in general and his *Principia* in particular in shaping both the nature of the scientific enterprise and our image of that enterprise. In this regard I am concerned both with Newton and with the use others have made of Newton, including uses of which he would disapprove, in the debate concerning the rationality of science.

The most obvious and direct fashion in which Newton influenced the scientific enterprise was through his role in creating the modern form of that enterprise. The *Principia* is no mere work within a tradition, it is a work which helped (to a large measure) to form a tradition within which scientists still work. The four most salient features of *Principia* in this regard are: fertility, axioms, mathematics, and explanation.

By "fertility" I mean to draw attention to the fact that *Principia* provides not a closed and finished theory but an open and ongoing research programme. In the preface, after drawing attention to the applications to be provided of his principles, Newton expresses the hope that "we could derive the rest of the phenomena of Nature by the same kind of reasoning from [these] mechanical principles." Even if history has not found that all phenomena can be encompassed within his theory, the extension, elaboration, and application of that theory provided the central focus for the activities of physicists until this century. So impressive was the Newtonian theory in this regard that there has been a convergence of opinion among philosophers of science ranging from Karl Popper to Thomas Kuhn that an essential feature of a good theory is its fertility. Newton and the subsequent developments of his theory are the best examples of this feature. In a fertile theory such as *Principia* the theoretical richness generates new and novel applications and developments.

The many theorems of *Principia* are derived from "axioms" with the help of definitions. Newton was not making an original move in casting his theory in the form of an axiomatized deductive theory. Since Euclid, philosophers, both natural and non-natural, had seen the axiomatic theory as the proper vehicle for organizing human knowledge. But Newton breathed new and lasting life into this form. In mathematics this approach needed no apologists. Outside mathematics the fruits of the method were at least contestable. His contemporary proponents of an axiomatic orientation, most notably Gottfried Leibniz and Spinoza, had struck Newton as barnacles on the ship of progress. And the growing empiricism of the times, with its rejection of the possibility of establishing substantial principles applicable to the empirical world by *a priori* means, might well have led to a distrust of the search for completely general truths which were to serve as the axioms. For such truths go far beyond the nitty-gritty of experience. The way forward, as the success of *Principia* established, was neither to proceed *a priori* nor to build slowly from experience. It was to produce, by conjecture, general axioms the vindication of which lay in the empirical successes they provided, successes at the level of predicting and explaining observed phenomena. Newton himself claimed quite unconvincingly to have derived his laws from experience. We learned how to do science from what he did, not from what he said.

It would be prudent, however, to sound a note of caution at this juncture. Unlike the matter of fertility, quite reasonable doubts have been articulated in recent years concerning the importance of axiomatic theories even in the physical sciences. For instance, it is claimed that in many cases the content of a theory in science is conveyed not by general axioms but by metaphors and/or models. In plate tectonics it is said that the Earth's crust is rather like a system of steel plates on a ball of plasticine. That model may be fertile, providing ideas to be explored about the formation of mountains and the cause of earthquakes, but the inherently metaphorical character of the model precludes any axiomatic representation of the theory it conveys. To force such theories into deductive forms would be to destroy them. And, indeed, it is said that the undue attention paid by philosophers to the giants, Newton and Einstein, has had a distorting impact on the philosophers' conception of science. But this may be merely sour grapes. Not having their own Newton these sciences seek to improve their status by deprecating the model his successes seem to vindicate. And in any event there is no sign whatsoever of a disenchantment with the axiomatic approach for fundamental physical theories.

As the work of the rationalists amply illustrates one can have axiomatics without mathematics. But one cannot have Newton's axioms without mathematics. Newton describes *Principia* as providing the "mathematical principles of philosophy" and describes the use made of geometry in his work as being to "the glory of geometry." The use of mathematics in physics was certainly not unique to Newton. In this case he did stand on the shoulders of giants, as he once remarked in an uncharacteristically modest moment. But the success of his mathematical principles meant that physics would be forever mathematical physics.

The system of *Principia* is more than a fertile, axiomatized system of mathematical physics. It is also an "explanatory" system. It invokes unobservable theoretical items in order to explain observable phenomena. Why does the water in Newton's bucket have a curved surface when the bucket and the water rotate together? Because the rotation of the water with respect to absolute space is producing a force effect. Absolute space is not something we can observe. It is something posited to explain aspects of the things we do observe.

The success of Newton's theories did much to render respectable the project of introducing reference to unobservable entities in the course of explaining the observable.

Fertility, axioms, mathematics, and explanations are the features of *Principia* in virtue of which it has come to provide the dominant model of a decent scientific theory. This description could suggest that it is merely a paradigm of the ideal theory. In fact what I wish to imply is something much stronger. The notion of the ideal physical theory is not some eternal Platonic form onto which Newton latched. Newton played a considerable role in creating the form. Scientific theorizing is a human activity which has a history. And in that history we have discovered not only particular theories, we have created a conception of what features make a good theory. Newton and his *Principia* represent an important stage in the evolution of the scientific enterprise.

Several references have been made to the success of Newton's *Principia*. The most visible and least contentious aspect of its success is the predictive power and its associated technological spin-off. Even if we had not got beyond Newton we could still have had the *Concorde*, the space programme and much else of what we regard (rightly or wrongly) as technological progress. That power to produce an impressive degree of mastery over nature, enhanced by the related understanding it provides of the physical world, played a crucial role in giving the scientific community the position and power it now has. We transfer vast resources to the institution of science and have come to accept an image of it as a very special institution with a particular claim to our allegiance. This image is the image of science as the very paradigm of institutionalized rationality. The continuing progress of science prompts the thought that scientists must have some special technique which they dispassionately and disinterestedly apply, taking us with each application a step closer to some noble (or Nobel) aim variously characterized as truth, explanation, or predictive power. Philosophers of science until relatively recently have accepted this image of science. And if they have offered a slightly more modest view of science, that is usually because they hold an even more modest view of other forms of human activity. Thus Popper writes: "The history of science, like the history of all human ideas, is a history of irresponsible dreams, of obstinacy, or error. But science is one of the very few human activities – perhaps the only one – in which errors are systematically criticized and fairly often, in time, corrected. ... in other fields there is change but rarely progress" (216). This image has issued in what I call "rational models of scientific change." Such a model involves two ingredients. First, one specifies something as the goal of science. It may be the discovery of explanatory truths or it may be the production of theories that are useful merely for making predictions. Second, some set of principles is specified for comparing rival theories against the evidence. The rationality of science is then displayed by showing that scientific change can be given a rational explanation by reference to the model. For example, in explaining why the scientific community changed its allegiance from Newton to Einstein at the beginning of this century one cites the following:

1) The scientific community had as its goal the goal posited by the model.

2) On the evidence then available, Einstein was superior to Newton as judged by the principles of comparison specified in the model.

3) The community perceived the superiority of Einstein over Newton.

4) This perception motivated the members of the community to abandon Newton for Einstein.

One would have to be touchingly naïve to think that this was all that was involved in a transition from one theory to another. Not even Popper is that naïve. Consequently those attracted to the rational image of science have drawn a distinction between internal and external factors which may influence scientific change. Internal factors are those which relate only to features of the theories, the relation between the theories, and the available evidence. All other factors are external. These would include psychological and sociological features of the scientists themselves and sociological features of their institutional context. When a given transition is amenable to a rational explanation, this type of external factor is irrelevant. These factors come into play only if there is a deviation from the norms implicit in the rational model. In that case the transition is turned over to the sociologists.

To illustrate, consider an example where the scientific community failed to change its allegiance as a rationalist might have expected. In the early 1800s Thomas Young articulated a wave theory of light which was largely ignored. In time the wave theory triumphed and it came to seem that that theory had been much better than the Newtonian-inspired corpuscularian theories even in the early 1800s. Those who see this as an episode of non-rationality have sought to explain the neglect of Young's theory by reference to the prevalent hero-worship of Newton, Young's poor manner of presentation, and an anonymous character assassination of Young published in the *Edinburgh Review*. Those who on the other hand want to maximize the amount of rationality in the history of science have sought to establish that Young's theory in its underdeveloped state was in fact inferior to the Newtonian theory given the evidence available at the time (Worrall).

What I have called the principles of comparison are perhaps more commonly referred to as a methodology. These are intended to give a systematic articulation of the scientifically respectable reasons for preferring one theory to another. Failing to give due accord to these reasons is variously described as failing to be scientific or failing to live up to the requirements of scientific rationality. The success of science has created the impression that this form of rationality has a special force or privileged claim on our allegiance. Failure to respect these norms is a failure to be scientific. It is to allow the intrusion of extra-scientific, non-rational factors.

This bipartite approach to the evolution of science is certainly attractive and has an initial plausibility. Science is seen as having its special method of which we are offered varying characterizations: John Stuart Mill's methods, Rudolf Carnap's inductive logic, Karl Popper's falsifications. When science is functioning properly we explain its transitions by displaying that scientists are making their decisions in accord with our favourite methodology. Ironically as it turns out, we have probably sought to vindicate it by reference to the work of Newton and Einstein. The special rationality of science is a matter of its having its own canons for rational belief and a special allegiance to accepting the verdict of those canons. If the community fails to proceed rationally we look for some external factor. For example, one finds members of the Academy of Science in East Berlin advocating Lamarckian theories of evolution. It is not absurd to argue that they are not living up to the norms of scientific rationality and to think that these theories are advocated not because they seem to the Academy to meet the canons of scientific method but because they actually meet the current ideological needs of the Communist Party.

Some philosophers of science such as Popper think that the external factors play only a very insignificant role in determining the evolution of science. Others who think that science does evolve rationally accord significant importance to external factors. And still others, the "non-rationalists," hold that external factors are the only important factors. For example, one such writer, Lewis Feuer, suggested that Einstein's theory of relativity triumphed because of the revolutionary atmosphere in Zurich: Lenin and Rosa Luxembourg were there at the time. Indeed Feuer goes so far as to claim that relativity theory could not have been successfully developed and propagated from Cambridge. In the context of this paper of particular interest is the notorious externalist account of the triumph of Newton by Boris Gessen, who purported to explain this by reference to the role the Newtonian theory played in serving the interests of the dominant social class. In addition to non-rational models which look to large-scale social and economic forces there are models which use games theory: these suggest that some scientists select theories according to which choice will be most likely to advance their careers (a trendy version of the old ploy of adopting the theory of the head of the department in the hope of getting tenure).

I cannot consider the details of various rational and non-rational models of science. My aim is the more general one of explaining how this controversy between the rationalists and the non-rationalists arose and to argue, with the help of Newton, against the simplistic terms in which it has been posed. In this regard there are two matters which need attention. First, if one proposes to treat differential scientific beliefs and belief transitions according to whether they are rational or not (cognitive in the one case, non-cognitive in the other), the way in which one distinguishes between what is rational and what is not becomes of the greatest importance. Second, one needs an account of whatever it is that could lead one to the puzzling and ultimately paradoxical conclusion that rational considerations play no role in the evolution of science.

I suggested above that the triumph and success of Newtonian science was a primary reason for the ascendancy of rational models of science. From the period of Newton until the end of the nineteenth century science seemed to fit a sort of building block model. There seemed to be a steady accumulation of truths, truths which provided the basis for impressive technological spin-offs. And, most importantly, the accumulation took place within a conceptual framework provided by Newton's theories of space, time, and mechanics. Of course there were anomalies such as the discrepancy between the predicted and the observed perihelion of Mercury. But these were conceived of as problems to be handled in due course within the Newtonian programme. And the success of that programme was appealed to in vindicating the accounts of scientific rationality. The rational image seemed triumphant.

The demise of the Newtonian world-picture in the early years of this century was probably the most traumatic event in the history of the philosophy of science. Newton held that to every action there is an equal and opposite reaction. The reaction, albeit somewhat delayed, by philosophers to this demise was opposite but far from equal. It was a classic case of over-reaction. The reaction took place in three areas. First, it prompted what I have elsewhere called the "pessimistic induction." The Newtonian theory which seemed so entrenched proved within less than 250 years to be false, strictly speaking. Of course this had also happened to Aristotle and to Galileo but the establishment and continuing success of modern science since the seventeenth century looked like the permanent basis for all future science. If even that basis could prove to be false in no more than 250 years, the prudent scientist should expect that even the most cherished theory will be disproved within 250 years. If this version of the pessimistic induction seems too pessimistic, one can replace the 250 years by five hundred years. The conclusion is the same. The evidence is that every theory will eventually turn out to be false. But then how can it be rational to do science? It would mean pursuing a goal which we have good reason to believe can never be realized. Not only will we have to take down some of our blocks and start again, we will have to build and re-build endlessly. Science comes to resemble more the fickle world of fashion than Popper's vision of steady progress. The reaction in the first case served to tarnish the traditional concept of truth as that which we were accumulating.

The reaction in the second instance relates to meaning. It seemed that, once one made the transition to the relativistic mechanics of Einstein, the very concepts in terms of which the Newtonian theory had been expressed were no longer available. Kuhn, for example, has argued in *The Structure of Scientific Revolutions* that the standard derivation of the Newtonian laws as limiting cases of the relativistic laws is fallacious. For, according to him, the notions defined as Newtonian within the Einsteinian framework are mere surrogates. If this is the case, questions arise as to how a Newtonian and an Einsteinian could communicate with each other. If there is no common language in which both theories can be expressed, there seems to be no possibility of any rational explanation of scientific change. It is as if I were asked to make a rational choice between two theories, one expressed in Serbo-Croatian and the other in Mandarin (I read neither of these languages). In Kuhn's vivid image (of doubtful intelligibility) the proponents of the competing theories inhabit different worlds. Even if, as I would argue, the situation is not so extreme as this, it remains the case that we have to cope with the fact that scientific change is not restricted to replacing old laws with new within the same conceptual framework. On occasion it amounts to the construction of a new framework and that creates complications, to say the very least, in making comparisons between the rival frameworks. It is this reaction that has done the most to create the impression that an external sociological treatment of scientific change is more appropriate than a rational, internal one.

The third area of reaction relates to the principles of theory choice. An awareness of the dramatic and extreme nature of scientific change prompted the thought that the very methods of science, the very principles of theory choice themselves, might change at the same time. The idea that there is a change in scientific methodology is one whose time had come. For there is something absurd in the picture of scientific method as something without a history. We had to create it and we did not at one particular time produce a recipe for use forever. And as I noted in the case of Newton, his *Principia* in part determined what was to be counted as a good theory. Kuhn suggested that the recipes were paradigm-bound. That being so there would be another reason for thinking that the rational assessment of rival theories was an impossibility. More recently, he has changed his mind (or has clarified what he meant all along) and argued that there are basic principles which articulate the good-making features of theories and which are common to all paradigms. Nonetheless there is room left for a substantial sociological account of theory transition in science. For Kuhn would claim that there is room for disagreement about the interpretation of the particular features in particular contexts and for disagreement about the weight to be given to different factors. A further source of doubt about the principles of theory choice which the rationalist assumes is in the difficulty of actually stating such principles. Indeed, Paul Feyerabend, in his aptly titled *Against Method*, argued that no exceptionless rules of any content can be found. And if there are no such rules, the prospects for a rational perspective are dim.

At this point the reader is likely to think "This is all crazy." To suggest that the pursuit of scientific truth is not rational or that scientists cannot communicate across rival theories and that even if they could they would have no rational grounds for preferring one theory to another is surely absurd. In the face of the success of science as measured by technological spin-off how could one possibly think this? Well some do: on two occasions I have debated on television with proponents of the sociological approach. Indeed the situation seems so absurd that one is apt to feel that what is needed is not so much a rational refutation of the position as a sociological or psychological explanation as to why anyone has ever believed it!

But first let us be clear about the basis for arguing against the conclusion which we are likely to agree is absurd. The premise to our argument is the undeniable claim that whatever else one thinks about science and rationality one has to admit that at the level of providing power over nature (as indicated by predictive power and technological spin-off) the history of science since the seventeenth century has been impressive (even, perhaps, frighteningly so). If the evolution of science which has produced this power was apparently explicable in terms of external factors one would generate something much more mysterious than what one was supposed to be explaining. Suppose theories were chosen to please the Church, the Party, or to enhance career prospects. In this case it would be utterly mysterious that choices made for those reasons should conspire together to produce the technological and manipulative power we witness today. The only plausible explanation of this phenomenon is the fact that the principles and procedures deployed are cognitively reliable, if fallible, indicators of the viability of theories. An attempt to give a purely external explanation of the continuing empirical successes of science fails for it generates a mystery greater than the one it aims to explain, namely, why should these external factors so nicely conspire to produce the empirical successes?

To refute the sceptic thus is no more satisfying than Dr Johnson's refutation of Bishop Berkeley by kicking a stone. A satisfactory answer must display the principles of theory choice and it must provide an account of how communication across conceptual shift is possible. And we can concede to our externalist the role that the sceptic provides in the history of epistemology. The sceptic prevents or should prevent us from becoming complacent. Without that challenge we might remain content with beliefs about the methods and goals of science that do not stand up to scrutiny. For example, until this challenge it was universally understood that the goal of science was truth. As a result of the challenge there has been fruitful and illuminating controversy among those of a rational orientation as to whether or not this is in fact so. Some now defend models of science in which the goal is not deeply explanatory truth but merely predictive power. Others have claimed that the goal is not truth *per se* but something more modest, the increase in the truth-content of theories. Embarrassingly, however, no one has come up with a convincing account of what truth-content is. If sceptics had not existed it would have been a good idea to have invented them.

The externalist, in articulating this untenable position, plays a useful role as a gadfly, prompting the rationalist to provide viable accounts of meaning, truth, and methodology which meet the objections noted above. But this is not a very satisfying role for mature philosophers and sociologists to play. What is it that has prompted them to do this? It cannot be very rewarding to have to articulate an obviously false theory for the sake of assisting the opposition to refine their own theories. Unless we want to offer a sociological or psychological account of the reasons some are inclined to play this role we must locate some rational basis for their reaction to the rational perspective. Perhaps this arises in part from profound disappointment in discovering that science is not a matter of steadily accumulating truths in the one true framework. After all, it is frequently said that there is no more militant an atheist than one raised in the faith. No doubt that is part of the explanation. But there is also the possibility that something is suspect in the position of the rationalist which prompts the over-reaction. This thought, to which I turn below, will take us back in an illuminating way to Newton.

I shall suggest that the non-rationalists, the externalists, have a point. Our cultural heritage provides us with a picture of scientific rationality as something privileged and regulative. The privileged character of science means that its results have a special claim on our allegiance. Its regulative character means that a non-cognitive attitude is taken to failures. But the externalists find that nothing corresponds to it. Their understandable dismay at this fraud leads them, naturally if mistakenly, to seek an account of science in which no such conception of rationality plays a part. This conception, which someone steeped in history or sociology is unlikely to see in the history and praxis of science, is the conception of the scientist with a special tool, the scientific method, which is unique in kind and which is deployed in the pursuit of the noble goal without deviation. That conception which runs through John Stuart Mill's methods probably reached its apotheosis in Popper and the positivists, some of whom actually believed in the existence of an algorithm for theory choice. The externalists reject that picture of rationality and rightly so, as I argue below.

It is a mistake, however, for the externalists to conclude that no conception of rationality is viable, for in that case their own position self-destructs. If there is nothing that counts as a rational reason for belief, if what we call *rational considerations* take their place on a par with bribes, threats, and prizes, we have no reason to take the externalists seriously. Once in a discussion with two externalists I asked why we were not simply paid to agree. They answered that it had not proved effective to treat academics in this fashion. Using the rhetoric of rationality was more effective. I would have preferred them to do a controlled experiment using escalating cash incentives for conversion.

One can illustrate nicely the difficulties in the conception of scientific rationality with Newton. Newton spent more time and energy on alchemy than he ever did on physics. And perhaps even more time on the interpretation of the scriptures. In the heyday of Newtonianism this work was conveniently forgotten. In 1880 the Portsmouth family which had inherited Newton's papers through his niece, offered them to the Cambridge University library, asking, *en passant*, that all papers without scientific value be returned. Some 121 boxes of papers on alchemy alone (in addition to those on theology) were returned. And Sir David Brewster, Newton's nineteenth-century biographer, on finding the "damaging" evidence of the papers on alchemy, could cope with this "problem" only by supposing that Newton was hoping to make new discoveries in chemistry through his study of alchemy. Maynard Keynes, who purchased many of the papers at auction, concluded that they were "wholly magical and wholly devoid of scientific value," and called Newton "the last of the magicians." Today, in this more enlightened age, these papers remain unpublished (Feyerabend).

This is very ironic. We have a picture of scientific rationality which ties it to certain procedures. Accounts of those procedures were vindicated by reference to the ways they fitted Newton in among the other scientific giants. But in this process of vindication attention was paid only to an aspect of Newton's overall procedure. And, having apparently vindicated their account, it is cited as a reason for rejecting much of his work as unscientific. Please do not misunderstand me. This is not a plea to bring back alchemy. It is part of my attempt to understand what drives the non-rationalists to their conclusions. A non-rationalist looking at Newton does not see a lack of continuity between his activities as an alchemist and his activities as a mechanical philosopher or a physicist. There is not sufficient difference to justify a non-cognitivist stance. Passing from *Principia* to the alchemical writings is not like changing from reading *Scientific American* to reading a Jehovah's Witness tract, a work on magic, astrology, or the speeches of Ronald Reagan, all of which give one the sense of being on holiday from rationality. There is here a seamless web. For in the alchemy, as in the physics, scrupulous attention is paid to carefully prepared and executed experiments with meticulous recording of the results. There is careful argumentation and reflection. There is the same aim of finding out the truth about the world. Externalists, seeing no discontinuity, cannot see why in the one case we take a cognitive stance, giving rational explanations, and in the other we take a non-cognitive stance, subjecting Newton to external treatment. They fail to see any special privileged techniques operative in the formulation of his physical beliefs that are not there in the case of his alchemical beliefs. There is not enough of a difference to rationalize providing respect in the one case (for the writer of *Principia*) and lesser treatment in the other (for the writer of the alchemy). And this attitude seems especially puzzling when you realize that the distinction between being scientifically rational and not being so is our distinction. It was not known to Newton and his contemporaries.

Perhaps Newton was a multiple schizophrenic, one moment the scientist, one moment the alchemist, one moment the interpreter of the scriptures. But in that case the mixed modes of his personality become very difficult to disentangle. This is so even apart from his alchemy. For it is there in the *Principia* and the *Opticks*. In the midst of what we count as his scientific works one finds theological considerations which certainly do not fit the rationalist picture of scientific rationality. Consider space and time: arguments are given for the existence of absolute space, a space which exists independently of the contents of space, based on the applications of his laws to the phenomenon of the water in the rotating bucket. But in the next breath other arguments are provided. God is infinite. Space and time are infinite. If they existed on their own there would be other infinite things besides God. This would detract from his glory. Hence, space and time must be attributes of God. God exists necessarily. Obviously, then, his attributes exist necessarily. It was a contingent matter that he created the world. It follows, or at least it does in the Latin, that space and time exist independently of their contents. Once upon a time, I remember with a shudder, in papers and lecture courses on the philosophy of physics I advised cavalierly that there were two Newtons, Good Newton and Bad Newton. That was absurd and presumptuous of me. But, and this reveals something about the prevailing picture, it did not strike people that way. As is (I hope) evident, I now find it unhelpful to describe Newton as not "scientific" or not "rational."

Of course, you will not get academic tenure these days in a physics department for pursuing his lines of thought on alchemy. But if you condemn the lines of thought implying that the explanation of why Newton should have advanced them must be an external one, then I object.

In considering Newton we can take the entire corpus of his work and measure it against some conception of scientific rationality. This will drive us to the absurd conclusion that on balance the greatest scientific genius of all time was not very scientific. Alternatively, we compartmentalize him. Part of his activities fit our rational model. In the other cases it does not and in these cases we deal with him psychologically, sociologically, or in terms of some lesser sort of rationality. But this the non-rationalists rightly see to be quite arbitrary and entirely unconvincing. If this is what using a conception of scientific rationality leads to, the externalist will have no part of it.

The solution, I suggest, is that we join the non-rationalist in jettisoning a conception of *scientific* rationality where this is intended to delimit some special and privileged form of rationality. Then we should look at Newton's alchemical and theological beliefs aiming to discover what reasons he would offer in their support. We should not ask as our first question whether some bit of the Newtonian corpus is scientific with a view to seeking a different treatment depending on the type of answer. We are likely to see reason at work and come to understand why he believed what he did through articulating the internal logic of his belief system. For example, Newton believed that God was revealed to us through his word and through his work. The former involved a study of the scriptures, the latter a study of physics. The scriptures held the promise that truth would be revealed to the pious man. And so it stood to reason that pious men of the past (including Paracelsus and the alchemists) had divined important things, things that would have been included in their works but in a coded form. The code hides the insights from the unfaithful. Newton's friend Fatio reports Newton as believing, for example, that Moses, Pythagoras, and Plato all knew the universal law of gravitation. And thus Newton had no alternative but to labour on their texts, to make extensive notes, to repeat their experiments. Newton had his reasons and we will understand him by seeing what they were. We will fail to understand him and fail to do him justice if we simply castigate these as non-scientific, requiring of them a non-cognitive treatment.

It did not work. There is no *Principia Alchemia*. It is a truth, albeit a sad one, that following the dictates of reason does not guarantee success. And ironically perhaps Newton put the final nail in the coffin of this sort of project. The fertility of his mechanics generated such excitement that interest in alchemy simply lapsed. I do not suggest it be revived. There are insuperable difficulties in the web of Newton's belief system. But we will never understand him if we put what is called "non-scientific" or "not rational from the scientific point of view" on the side for some sort of "other" treatment.

Rationality in matters of belief and action is a good thing. It is in the case of belief a matter of distributing one's degree of belief with an eye to the evidence. But at any given moment our interpretation of the evidence takes place against our background beliefs which will include beliefs we sometimes find convenient to classify (somewhat arbitrarily) as metaphysical, theological, scientific, and so on. There is no such thing as scientific rationality. Of course we can speak of such a thing in a harmless way if we wish, as a way of talking about scientists being rational. Their being rational consists in their being the way we are when we are rational. Neither Newton nor contemporary scientists have any special mode of rationality. On this broader conception of rationality we have less to turn over to the psychologist or the sociologist for much more will turn out to be rational even though it may transpire that we do not call it scientific.

The non-rationalists fail to see a discontinuity where indeed there is none. Their mistake is to reject the concept of rationality and rational explanation rather than reject the conception of *scientific* rationality. I am not saying that there is nothing special about science. Obviously there is. But what is special is not the scientific method, a special form of rationality. All the "specialness" we need can be found in the norms of the institution of science. Those norms include the requirement of public debate, of the articulation of reasons, and of the search for explanatory theories with predictive power of sensitivity. In this, science contrasts with institutions of a social or political character. The aim is often to arrive at a consensus on what is to be done. We all recognize that that aim may best be served on occasion by not saying certain things. A political figure who is being "economical with the truth," in Sir Robert Armstrong's memorable phrase (used in the course of the attempt by the British government to have Peter Wright's *Spycatcher* banned) is being a typical politician. The norms of the scientific institution mean that an experimenter economical with the truth when it comes to adverse results is likely to find himself removed from the community.

We should approach the history of science with the expectation that it can be explained largely in rational terms. The success of science is what justifies our presumption of rationality. But it is not appropriate to assume that this can be done using some particular model of scientific rationality. We can talk of rationality and look to the scientists' reasons for their beliefs and actions without being too concerned as to whether those reasons meet some special canons of scientific rationality. As noted above, the idea of a rational model of science has an initial plausibility. But in the end it has a distorting effect in that it has inclined some (including the author, in a previous work) to leave too much of the work of figures such as Newton for external treatment. Using a broad and general notion of rationality we come to a better understanding of why Newton held the views he did, an understanding which will leave less to be treated externally.

And if Newton failed, as even the greatest do, it was not because he failed to follow the scientific procedures in his alchemical studies or forgot the standards of rationality. It was because he failed to test his alchemical beliefs in the public forum of the Royal Society. And that may well need an externalist story. Newton was a complicated, difficult, and strange man. No doubt the death of his father before his birth and the subsequent remarriage of his mother played a role in creating that personality. It was the personality of one who did not want to go public unless he was confident of triumphing. He had not made the requisite progress which would have given this confidence in his non-mechanical studies, so he sent the Royal Society his telescope, not his alchemical writings.

To return to my marchioness. It is a fancy of mine that she understood what I am trying to say. She took Newton's theory and made of it what she could. She debated its application with the narrator, objecting that gallantry is not a relevant consideration. Offered the alchemical or theological results I see her proceeding gaily in the same way without turning them aside as unscientific, without condemning them as being beyond the pale because they were not scientifically rational. She would not see them as requiring explanation by reference to the psychology of Newton nor the sociology of the times. Newton's reasons for holding them would provide the explanation of why he held them and provide the basis for assessing whether she too should hold them or reject them. No doubt the marchioness would have terrified Newton and she might well have felt the need of some psychological accounting for that fact. But that is another story.

I have projected a vision of Newton as one of the creators of the modern form of the scientific enterprise through certain characteristics of his *Principia*. The fame of that success played a crucial role in the development of an image of science with a special form of rationality. That form of rationality, I have suggested, is a mystic structure which needs to be dissipated. Unless it is dissipated and replaced by a richer, less partisan, conception we will never understand Newton. If you think that this conception, which I challenge, is not operative, ask yourself why there is no edition of Newton's alchemistry currently available. But we must not follow the non-rationalists who in over-reaction jettison the conception of rationality entirely in self-destructing fashion. Without a conception of rationality and the assumption that we do to some degree follow the dictates of reason we will never understand the success of science. Of course, dissipating this may leave us with a conception of science which is nothing more than a convenient category. But that may be no bad thing, particularly at this moment in time. I sense, at least in the United Kingdom, the deployment of the myth of scientific rationality afresh. Various physicists are making *ex cathedra* pronouncements about what one should believe when it comes to certain questions about God and metaphysics. I do not say that they thereby fail to be scientific or to be rational. But I do say that we ought to look at the merits of the case advanced without presuming that it has any special merit merely because it has the backing of those paid to do science. But the last word must go to the one whose practice was invariably to claim the last word. The quotation from his alchemical writings reveals his grandiose hopes for that subject.

> For Alchemy tradeth not with metalls as ignorant vulgars think, which error hath made them distress that noble science; ... . This Philosophy is not of that kind which tendeth to vanity & deceipt but rather to profit & to edification inducing first ye knowledge of God & secondly ye way to find out true medicines in ye creaters ... ye scope is to glorify God in his wonderful works, to teach a man how to live well. ... This Philosophy both speculative & active is not only to be found in ye volume of nature but also in ye sacred scriptures, as in Genesis, Job, Psalms, Isaiah & others. In ye knowledge of this Philosophy God made Solomon ye greatest philosopher in ye world. (qtd. in Christianson 223)

Such were Newton's hopes for alchemy, which it is to be noted include the discovery of medicines and the increase in our knowledge of God. In the seamless web of his activities, Newton's hopes for mechanics were not less grandiose: "the mechanics manifested in natural phenomena served as windows through which man could catch fleeting and imperfect glimpses of the higher reality and ultimate purpose of creation" (qtd. in Christianson 224).

WORKS CITED

Algarotti, Francesco. *Sir Isaac Newton's Philosophy Explained for the Use of the Ladies*. London, 1739.

Christianson, Gale E. *In the Presence of the Creator: Isaac Newton and His Times*. New York: Free Press, 1984.

Dobbs, B.J.T. *Foundations of Newton's Alchemy*. Cambridge: Cambridge UP, 1975.

Feuer, Lewis S. "The Social Roots of Einstein's Theory of Relativity," in *Einstein and the Generations of Science*. New York: Basic Books, 1974.

Feyerabend, Paul K. *Against Method*. London: New Left Books, 1975.

Gessen, Boris Mikhailovich. *The Social and Economic Roots of Newton's* Principia. New York: H. Fertig, 1971.
Kuhn, Thomas. *The Structure of Scientific Revolutions*. 2nd ed. Chicago: Chicago UP, 1970.
Newton-Smith, William H. *The Rationality of Science*. London: Routledge and Kegan Paul, 1963.
Popper, Karl R. *Conjectures and Refutations*. London: Routledge and Kegan Paul, 1963.
Westfall, Richard S. *Never at Rest: A Biography of Isaac Newton*. Cambridge: Cambridge UP, 1980.
Worrall, J. "Thomas Young and The 'Refutation' of Newtonian Optics: A Case Study in the Interaction of Philosophy of Science and History of Science." *Method and Appraisal in the Physical Sciences*. Ed. C. Howson. Cambridge: Cambridge UP, 1970. 107-79.

Newton and Adam Smith

D.D. RAPHAEL

Professor Richard Westfall, in the first of these papers, has said of Isaac Newton's *Principia* that no subsequent event, to his knowledge, has matched the impact on western civilization of the publication of that work in 1687. The nineteenth-century historian, Henry Thomas Buckle, said of Adam Smith's *Wealth of Nations*: "looking at its ultimate results, [it] is probably the most important book that has ever been written" (1: 194). Both these judgments reflect the enthusiasm of the partisan, but that of Professor Westfall (who candidly acknowledges that he is "not exactly an impartial judge") is both more sober and more worthy of acceptance: Adam Smith did not possess the towering genius of Newton and his book did not represent a climax of fundamental discovery. Yet one can reasonably say that the *Wealth of Nations* is the greatest classic work in the social sciences, as the *Principia* is in the natural sciences, and it is worthwhile to consider how far Smith's achievement was influenced by Newton's.

It may seem strange at first sight to speak of a book on economics being influenced by a book on physics. We can, however, find the connection in the work of Adam Smith himself, since he wrote a long essay on "The History of Astronomy" which shows a sound acquaintance with the contribution of Newton and which also sets out a general theory of scientific inquiry that can be applied to Smith's own achievements in economics and in moral philosophy. Writing elsewhere about scientific method, Smith praises Newton's method of explanation as the one to be followed in any scientific endeavour. We should not suppose that the scientific aspects of Adam Smith's work are confined to what he wrote about the natural sciences, namely the essay on "The History of Astronomy" and a related fragment on "The History of Ancient Physics." Today we worry ourselves with the question whether economics and other disciplines in the field of social studies should or should not be called sciences. We are quite right to debate the question, for it involves important methodological issues about the character of social studies. The picture appeared differently to scholars of the eighteenth century. For them the progress of physics, culminating in the work of Newton, was the paradigm of success in intellectual inquiry, and many of them aimed at emulating the progress of physics in other fields of inquiry, which they called, more or less indifferently, "philosophy" or "science." This was certainly true of Adam Smith. He saw his treatise on economics as a work of "science" or "philosophy," a scientific "system" comparable to the "systems" or theories of astronomy that he had described in his historical essay. The book is in fact more than that, since it also includes policy recommendations and lengthy digressions on historical or sociological topics. Nevertheless, the fact remains that the most important part of the work, the first half, consisting of Books I-III, which lives up to the full title of *An Inquiry into the Nature and Causes of the Wealth of Nations*, is notable for its scientific character; it is a systematic explanation of a diverse set of phenomena by means of an integrated theory. Although the individual parts of the theory are not especially original, their integration into a single, interdependent system was profoundly original and is the reason why Smith's book has pride of place as the primary classic work of the social sciences. It seems to me that Smith's achievement in this regard was undoubtedly influenced by the example of Isaac Newton.

Advance in science and philosophy alike has often been the result of adopting an idea or method that has been successful in one field and of transferring it to another. The overwhelming success of Newton in the study of nature led many inquirers to think of following his example in the study of man. Needless to say, their understanding of Newton's work was usually superficial - not surprisingly, when several heads of Cambridge Colleges in Newton's own day had said of the *Principia* "that they might study seven years, before they understood anything of it"(qtd. in Westfall, *Never at Rest* 468). There were a few non-scientists who had a reasonable grasp of Newton's natural philosophy, and not only among metaphysical philosophers like Locke and Berkeley. Samuel Clarke, a leading theologian and moral philosopher of the early eighteenth century, could be trusted by Newton to write on his behalf in correspondence with Leibniz about the concept of absolute space and related matters; Voltaire's popularization of Newton's scientific work is not negligible; and Adam Smith's account of Newton's place in the history of astronomy shows accurate scientific knowledge as well as acute philosophical insight. But none of these thinkers consciously aspired to do for the human sciences what Newton had done for the physical.

That aspiration was voiced by Jeremy Bentham, a thinker who made a valuable contribution to the philosophy of law but who is best known as the leader of the nineteenth-century utilitarian movement. Utilitarianism is the moral theory which says that actions and governmental policies are right if they aim at the greatest possible happiness for the greatest possible number. This is an important theory, which has the virtue of connecting together in a single, simple formula the fields of ethics, politics, and law. Bentham thought it was scientific in being built upon empirical facts. He also tried to make it more scientific by introducing a supposedly mathematical schema for calculating the "sum" of the various "dimensions" of pleasure or pain in the consequences of an action. At an early stage of his work on legislation (in the latter part of the eighteenth century), Bentham wrote of himself as consciously imitating the method of the physical sciences. He compared an earlier utilitarian, Helvétius, with Francis Bacon, the founder of inductive logic, and then looked for a comparison with Newton:

> The present work as well as any other work of mine that has been or will be published on the subject of legislation or any other branch of moral science is an attempt to extend the empirical method of reasoning from the physical branch to the moral. What Bacon was to the physical world, Helvétius was to the moral. The moral world has therefore had its Bacon, but its Newton is yet to come. (qtd. in Halévy, *La Formation* 1: 289-90)

He presumably meant that Helvétius had described in general terms the right method for a scientific treatment of ethics but there was still the more important task of actually applying the method and working out the details. Bentham was not a modest man, and although he did not name the Newton yet to come it was obvious whom he meant.

Bentham may have been led to make the comparison with Bacon and Newton by having seen it used of Montesquieu and Adam Smith in the field of historical sociology. A distinguished pupil of Adam Smith, John Millar, writing in 1787 about Smith's "lectures on the History of Civil Society" (meaning the historical part of his Lectures on Jurisprudence), said: "The great Montesquieu pointed out the road. He was the Lord Bacon in this branch of philosophy. Dr. Smith is the Newton" (*EPS* 275*n*).

Bentham's predecessors themselves, however, did not have his degree of self-assurance and indeed thought of themselves as disciples rather than emulators of Newton. Francis Hutcheson, professor of moral philosophy at Glasgow in the early part of the eighteenth century, propounded a theory of ethics which reduced all virtue to the motive of benevolence. In a way this is just a secularized version of the Christian doctrine that love encompasses all the virtues. Most philosophers have thought that such a scheme oversimplifies the facts. Hutcheson was quite ingenious in trying to show that apparently distinct virtues, like honesty, courage, and justice, are at bottom species of benevolence because their ultimate aim is to benefit other people or society at large, which, of course, is precisely what benevolence aims at. Hutcheson also noted that our strong love for those who are near and dear is only a small part of human benevolence, which extends beyond family and friends to fellow countrymen, to humanity at large, and even beyond the human race to sensitive creatures generally. Having explained that all virtues are united in the one virtue of benevolence and that it extends its influence over the whole of mankind so as to bind them together in a single body or system, Hutcheson compared the role of benevolence in ethics with that of gravitation in physics. The analogy was plainly prompted by a layman's superficial acquaintance with Newton.

> This UNIVERSAL BENEVOLENCE toward all men, we may compare to that principle of GRAVITATION, which perhaps extends to all bodies in the universe; but increases as the distance is diminished, and is strongest when bodies come to touch each other. Now this increase of attraction upon nearer approach, is as necessary to the frame of the universe, as that there should be any attraction at all. For a general attraction, equal in all distances, would by the contrariety of such multitudes of equal forces, put an end to all regularity of motion, and perhaps stop it altogether. (Hutcheson, V.2)

Like Bentham after him, Hutcheson also tried to make his theory more scientific by subjecting it to mathematical expression. His equations and theorems were indeed worked out with greater precision than Bentham's so-called calculus, but they had no practical value for all that, and when a friendly critic pointed out that the theory gained no benefit from them, Hutcheson had the good sense to omit them in a revised edition of his work. The long title of the original version had included the words "with an attempt to introduce a mathematical calculation in subjects of morality," indicating that Hutcheson initially attached much importance to his equations and perhaps indicating also a conscious desire to follow, in small measure, the example set by Newton's *Principia*.

The words that I have quoted from Hutcheson's title find an echo, surely deliberate, in the subtitle of a vastly superior work of Scottish philosophy, published fourteen to fifteen years later, David Hume's *Treatise of Human Nature*: "being an attempt to introduce the experimental method of reasoning into moral subjects." Hume was no mathematician and knew it. He did not try to invent any fancy mathematical formulae, but he did think, at first, that he was following a Newtonian pattern. When he used the word "experimental" he was not referring to experiment in the modern sense, the deliberate setting up of conditions for making an observation which cannot be made by just looking at ordinary natural events; by "the experimental method of reasoning" Hume meant what Bentham meant (and indeed it seems likely that Bentham was echoing Hume with his "attempt to extend the empirical method of reasoning from the physical branch to the moral"); he meant an empirical method, reasoning based upon the observations of experience instead of upon purely abstract principles. In itself, of course, that had no particular reference to Newton, but the influence of Newton's theory of gravitation can be seen, as it can in Hutcheson, when Hume based his psychological explanations on the association of ideas. "Here is a kind of ATTRACTION, which in the mental world will be found to have as extraordinary effects as in the natural, and to shew itself in as many and as various forms" (Hume I.i.4). This sounds more general than Hutcheson's analogy between gravitation and benevolence. In fact, however, Hume's analogy has more point, since he thinks of ideas as if they were separable particles and he sets out the principles of association in terms of specific laws so as to explain the mental processes. In an *Abstract* of the major, epistemological, parts of the *Treatise*, Hume highlighted his theory of the association of ideas as a discovery that might entitle him to the name of "an INVENTOR."

It has in the past been fashionable to say that Hume's aim in the *Treatise* was to become "the Newton of the moral sciences." One or two recent commentators have been sceptical, claiming that there were stronger influences on Hume from other quarters. I think it is more true to say that Hume did at first hope to follow in the path of Newton but then lost his zest for such a plan, partly because of difficulties that arose for his associationist account of the self and partly through disillusion with Newton's attempt to tie natural philosophy to natural religion.[1] Newton's contribution to theology is simply a curiosity now, but he himself took it very seriously at the time. So did eminent theologians like Samuel Clarke and Richard Price, who shared Newton's adherence to Arianism and his rejection of the orthodox Athanasian view of the Trinity. Newton reached his theological views from a detailed study of biblical and patristic texts (Westfall, *Never at Rest* 309-30), but the *Principia* itself contains a significant monument to his theological interests. One criticism of the original version that greatly irritated Newton was the complaint that gravity was an occult quality, going beyond principles of mechanics. In adding a concluding General Scholium to the revised second edition of the book, Newton replied to this criticism by connecting natural philosophy with theology by means of the Argument from Design. His example was followed by others, including Colin Maclaurin, who introduced the teaching of Newtonian physics into the University of Edinburgh at a time when Hume was a student there. One of the signal achievements of Hume's philosophy was to demonstrate the shortcomings of the Argument from Design as the stronghold of natural religion. Its use by Newton and the Newtonians seems to have strengthened Hume's disillusion with his initial aim of constructing a Newtonian science of man.

The influence of Newton on Adam Smith went deeper. I have no wish to exaggerate it and I must emphasize that it fell short of the influence of Hutcheson, who was Smith's teacher at Glasgow University, and that it came nowhere near the influence of Hume, whom Smith regarded as his closest friend and as the foremost philosopher of the age. Nevertheless the influence of Newton was real and substantial, simply because Smith was able to read and follow the *Principia*, at least in its account of astronomy. To judge from his published work, Smith's turn of mind was not instinctively mathematical, but he did have a degree of competence in mathematics that was lacking in Locke, Hutcheson, and Hume alike. His first biographer reported, from sound sources, that when Smith was an undergraduate student at Glasgow his favourite subjects were mathematics and physics, and that he showed particular ability in geometry.[2] The late Dr W.P.D. Wightman, when editing Smith's essay on "The History of Astronomy," acknowledged the essential accuracy of its account of Newton's astronomy but doubted whether Smith ever "studied" the *Principia* because he wrote as if Newton's force of attraction presented no problems. Wightman therefore suggested that Smith might have acquired a general knowledge of Newton's system of astronomy from Voltaire's *Eléments de la philosophie de Newton* or from Maclaurin's *Account of Sir Isaac Newton's Philosophical Discoveries* (Smith, *EPS* 21). This view, however, sits ill with the evidence of Smith's personal library, which included Newton's *Arithmetica Universalis*, *De Methodis Serierum et Fluxionum*, *Opticks* (both English and Latin versions), and *Principia*. Of course, a man does not necessarily read all the books he has bought for his library. Nevertheless, while Smith may not have "studied" the *Principia* closely, it seems to me unlikely that he did not read a fair amount of the book for himself.

This conclusion is supported by attention to a particular paragraph in Smith's account of Newton's system. There is good reason to think that the major part of Smith's essay on "The History of Astronomy" was written in his youth and that the section on Newton was added some years later. The writing of the Newton section can in fact be located between two precise dates, 1749 and 1758. It must have been before 1758 because Smith says that Newton's "followers have, from his principles, ventured even to predict the returns of several [comets], particularly of one which is to make its appearance in 1758."[3] The final clause refers to Halley's Comet. Smith's essay was published posthumously by his literary executors, and it was left for them to add a note saying that the essay was written before the date mentioned by Smith and that the comet did in fact return in that year, as had been predicted. The earlier date of 1749 can be fixed because Smith writes of observations in Lapland and Peru which confirmed Newton's deduction and calculation of a flattening of the Earth at the poles. The results of the observations to which Smith is referring were published in

1738 for Lapland and in 1749 for Peru. Some further explanation will show why this particular paragraph[4] of Smith's essay indicates that he probably did read parts of the *Principia* for himself and with attention.

Ever since Pythagoras astronomers had taken it for granted that the Earth was spherical in shape. Newton in the *Principia*, and Huygens shortly afterwards, showed that it must be an oblate spheroid, that is to say, that it departs from the spherical in being elongated at the equator and flattened at the poles. This is a natural effect of the rotation of a planet round its axis. Newton calculated the difference as in the proportion 230:229. He went on to show that it would be much greater for the planet Jupiter owing to its greater size and the higher speed of its rotation. His calculation of the degree of flattening in Jupiter was confirmed by telescopic observations in 1719. Having demonstrated the general theory, in Proposition 19 (Problem 3) of *Principia*, Book III, Newton proceeded, in Proposition 20 (Problem 4) to explain a phenomenon that had puzzled the scientists of the time. In 1672 a French astronomer, Jean Richer, had found, on an expedition to the island of Cayenne, near the equator in South America, that the pendulum of his clock was moving more slowly than it had done in Paris, and that it needed to be shortened by one-twelfth of an inch. This phenomenon was confirmed by later observations in Africa and again in America. Newton showed that it could be explained by the weaker effect of gravity near the equator if the equator was further from the centre of the Earth than was the latitude of Paris. Huygens reached similar conclusions in his book *De Causa Gravitatis*, published in 1690.

But this advance in scientific understanding received a check in 1720, when Jean-Dominique Cassini and his son Jacques measured arcs north and south of Paris, the results of which implied the reverse of the conclusions reached by Newton and Huygens. The Cassini measurements indicated that the Earth is flattened at the equator and is elongated at the poles. A repetition of their exercise in 1740 showed that their measurements were incorrect. Meanwhile, however, the French Academy of Sciences had decided to obtain more significant observations by sending an expedition in 1735 to a region of Peru near the equator and another in 1736 to a region of Lapland not far from the Arctic Circle. The Lapland expedition returned in 1737 and its findings were published in a book by Maupertuis in 1738. The Peruvian exercise was more elaborate and gave rise to conflicting interpretations. That expedition returned in 1743 and its findings were given in two books, one written by Pierre Bouguer and published in 1749, the other written by Charles-Marie de La Condamine and published in 1751. The upshot of the two expeditions was to confirm the theory of Newton and Huygens and to provide a calculation of flattening which was fairly near the figure reached by Newton. (Newton's figure is in fact nearer still to the one now accepted as the true value.)

While Adam Smith does not give the precise historical references, his account of the matter is detailed and almost entirely accurate. The only lapse is an apparent suggestion that there had been more than the one pair of empirical measurements casting doubt on Newton's theory. It is also worth noting that Smith had in his personal library a copy of Maupertuis's book on the expedition to Lapland. In the light of all this, it seems to me inconceivable that Smith would not have read for himself the two propositions as set out in his copy of the *Principia*.

Smith will not have given this degree of attention to more than a few parts of the *Principia*. The section on "The Figure of the Earth" will have interested him particularly because it is a striking example of the relation of theory to empirical evidence in brilliant scientific discovery. Smith's concern in his essay was with the history of astronomy and its implications for the philosophy of science. There is a passage in his Lectures on Rhetoric where Smith refers to Newton's method of scientific explanation, contrasting it with that of Aristotle.

> [I]n Natural Philosophy, or any other science of that sort, we may either, like Aristotle, go over the different branches in the order they happen to cast up to us, giving a principle, commonly a new one, for every phenomenon; or, in the manner of Sir Isaac Newton, we may lay down certain principles, known or proved in the beginning, from whence we account for the several phenomena, connecting all together in the same chain. The latter, which we may call the Newtonian method, is undoubtedly the most philosophical, and in every science, whether of Morals or Natural Philosophy, etc., is vastly more ingenious and for that reason more engaging than the other. (Smith, *LRBL* 145-6)

In these lectures Smith is chiefly concerned with questions of aesthetics and of effectiveness in exposition, but he does say here that the Newtonian method is not only "more engaging" but also "the most philosophical," meaning that it is superior as an explanation. Since the quotation comes from lectures that Smith was giving at Glasgow University alongside his regular course on moral philosophy, he must have thought that he himself followed the Newtonian method in his own discussion of moral philosophy (which included his economic theory).

Smith was well aware that the superior method was not invented by Newton (he himself says that it began with Descartes) but he was evidently most impressed by Newton's use of it, presumably in the *Principia*, where a vast range of phenomena is explained by a relatively small number of general principles. In the passage that I have quoted, Smith recommends the Newtonian method for "every science, whether of Morals or Natural Philosophy, etc." Examination of his two books, on ethics and economics, discloses what he had in mind as the Newtonian method. He begins with what he takes to be the basic cause of the phenomenon he intends to explain, and builds up from it in a systematic pattern. *The Theory of Moral Sentiments*, Smith's first book, begins with a chapter on Sympathy, because Smith regards that as the basic element of moral judgment, which is the main subject of the book. *The Wealth of Nations* begins with a chapter on the Division of Labour, because Smith regards that as the basic cause of economic growth, which is the main subject of this book. The two works share a symmetry of structure, including the elimination of alternative theories in the course of a historical survey of them after Smith's own theory has been expounded. This last feature is not a necessary part of "the Newtonian method"; it is a consequence of Adam Smith's own method of making headway in an inquiry. But the whole idea of a structure proceeding by connected stages, which underlies both of Smith's books, is clearly due to his notion of the Newtonian method of explanation. You can see what a difference it makes if you compare the finished form of *The Wealth of Nations* with the earlier version of Smith's economic thought given in his Lectures on Jurisprudence. Part of the superior power of *The Wealth of Nations* is due to Smith's use of colourful phrases for key concepts, but part is also due to the cumulative effect of a progressive structure which is not obtrusive but is perhaps more persuasive for just that reason.

Within *The Wealth of Nations* we soon find the same sort of Newtonian imagery that Hutcheson and Hume had used. Having explained the primary role of the division of labour, of a market, and of money, Smith proceeds to his theory of value, beginning with prices. He distinguishes between market prices and what he calls the natural price, and he uses the metaphor of gravitation to describe the relation between them.

> The natural price, therefore, is, as it were, the central price to which the prices of all commodities are continually gravitating. Different accidents may sometimes keep them suspended a good deal above it, and sometimes force them down even somewhat below it. But whatever may be the obstacles which hinder them from settling in this center of repose and continuance, they are constantly tending towards it. (Smith, *WN* 75)

The image recalls Hutcheson's comparison between universal benevolence and gravitation. But although less elaborate, Smith's comparison is more realistic in that he is dealing with phenomena that are already quantified. Market prices are the actual prices charged. The market prices for a particular commodity at different times can be easily plotted so as to show how they cluster round a particular figure over a period of time. There is, of course, no suggestion that this apparent centre of "gravitation" exerts any force on the actual market prices. The forces which determine prices are supply and demand in relation to the simple human desire to obtain what one seeks for as little sacrifice as possible. What is significant, however, is the possibility of working out a mathematical relationship. The so-called natural price is as much a fiction as the metaphorical idea of gravitational attraction, but the imagery enables the reader to see more clearly the kind of pattern that market prices display over a period of time.

Other notions of equilibrium in Smith's economic system should be understood similarly. He describes an equilibrium between the "toil and trouble" of work and the compensating enjoyment of pay; and another between the disadvantages of an unpleasant or risky business and the compensating advantages of high profits. In all this Smith has at the back of his mind an analogy with a system of equilibrium in physics, and it is likely that he is thinking in particular of the solar system as described by Newton. But we must not suppose that he carried the analogy so far as to think that the system was sustained by quasi-gravitational forces. Smith regarded his "system," and indeed all scientific systems, as products of the human imagination. They are methods of enabling us to synthesize the observable facts, but are not necessarily true representations of what exists in nature.

Smith's more famous image of "an invisible hand" has the same sort of function as his metaphors of gravitational force. In *The Theory of Moral Sentiments* he says that when the wealthy employ lots of servants and helpers they are unintentionally led by an invisible hand to reduce the inequality between rich and poor; and in *The Wealth of Nations* he says that individuals who intend only their own interest in their economic activities are led by an invisible hand to promote the interest of society. These passages are not statements of theological belief; they simply use an imagined fiction to help us to see effects which come about without being aimed at.

I come back to Smith's essay on "The History of Astronomy." Although the essay is impressive as an exercise in the history of science, Smith's purpose in writing it was to go beyond history to philosophical theory. This is characteristic of his approach to any large-scale piece of work. He begins with history and hopes to elicit from it a philosophical theory to explain the historical facts. That is how *The Wealth of Nations*, a theoretical treatise on political economy, arose from lectures on the history of law and government. It is how *The Theory of Moral Sentiments*, a theoretical treatise on moral judgment, arose from lectures which began with the history of ideas about morality. It is also how Smith developed a theory in the philosophy of science from investigations into the history of astronomy and of physics. The essay on "The History of Astronomy" is the first of a planned trilogy of essays, the other two dealing with the history of ancient physics and the history of ancient logic and metaphysics. The title of each essay states that the history is intended to "illustrate," provide the evidence for, a general thesis, "The Principles which lead and direct Philosophical Enquiries."

The historical course of astronomical inquiry is a particularly useful example because of the striking differences between theories, all of which were constructed to explain the same observed facts. The switch from the Ptolemaic to the Copernican theory is the most radical of these changes (Smith was, not surprisingly, unaware that a heliocentric theory had been proposed in ancient times, too, by Aristarchus of Samos); but other switches, well before Ptolemy and well after Copernicus, had also meant a fairly profound shift of ideas, as when the theory of concentric heavenly spheres was replaced by the theory of eccentric spheres, and when Descartes's theory of vortices was replaced by Newton's theory of gravitation. From reflection upon these differences and the shake-up of psychological attitudes which they implied, Smith produced a theory about the nature of scientific explanation which was itself, in its own way, quite radical too.

It states that the unobserved entities or forces which scientific theory uses to explain observed phenomena are the products of human imagination. Smith begins with the old Aristotelian remark that "philosophy," rational inquiry, originates in wonder. He interprets this as surprise at the unfamiliar, an uneasiness which is removed by finding that the unfamiliar can after all be shown to be an example of the familiar. For this purpose you need to fill in the gaps in your observation. You see the sun, the moon, the stars, up there in the sky. Some of them are observed to move, others to remain fixed. How do they do it? Why, in the first place, do they stay up there in the sky? And then, secondly, why do the sun, the moon, and some of the stars, the "wandering ones," move, while the rest of the stars stay put? The surprise is allayed if you imagine them as attached to transparent solid spheres surrounding the Earth. The sphere to which the sun is attached moves round the Earth regularly in a period of twenty-four hours. The sphere to which the moon is attached also moves but in a more complicated way. So do the separate spheres of the planets. But the outermost sphere, to which the other stars are attached, remains fixed. You cannot see these spheres, but you imagine them. When later investigators find that system too simple to accommodate the observed phenomena, they construct an alternative system (or "imaginary machine," as Smith calls it). The imagination supplies entities which are missing from the observed phenomena but which serve to remove the strangeness of the observed phenomena and make them seem not so strange after all.

Smith derived his view of the role of imagination from David Hume's theory of the external world. Hume's theory maintains that, while our perceptions consist of fleeting and fragmentary impressions, we obtain the idea of permanent objects out there because our imagination fills in the gaps between impressions. For example, I look out of the window and observe a cat sitting on the wall. I bring my gaze back to my desk. A few seconds later I look out of the window again and observe the cat on the ground. My imagination fills in the gap with an idea of intermediate perceptions of the cat jumping down from the wall to the ground. Otherwise I would be in the position of Alice in Wonderland, who finds the Cheshire Cat suddenly appearing and disappearing in different places with no continuity to assure her of a permanently existing animal. I do not know whether Smith accepted his friend Hume's theory of the external world. He may have done so in his youth, when he first read Hume's *Treatise* and when he could himself write an essay "Of the External Senses" which owed much to Berkeley. In his later life Smith had no appetite for metaphysics and seems to have been much more ready than Hume to accept the views of common sense. But scientific theory, once it has got beyond the elementary stage, is not common sense. Common sense will tell you that the sun moves, since it is apparently seen to do so, but common sense will not tell you that the sun stays suspended in the sky because it is attached to a crystalline sphere. So Hume's theory, that the imagination fills in gaps between observed impressions so as to produce the idea of solid bodies, seemed to Smith to be more plainly applicable to scientific theory than to our everyday view of familiar material things.

It follows from Smith's philosophical theory of scientific explanation that no scientific theory can be taken to be absolutely true. He notes this consequence himself at the end of his essay on "The History of Astronomy," after he has explained why the theory or system of Newton is superior to the theory of Descartes, which had been accepted by a large number of scientists for quite a long time. Smith then ends the essay by reminding the reader, and himself, of his thesis that all "philosophical" (including scientific) systems are "inventions of the imagination." The Newtonian system is so persuasive, he says, that he has been "insensibly drawn in" to write as if its "connecting principles" were "the real chains which Nature makes use of to bind together her several operations" (Smith, *EPS* 105). He clearly finds it difficult to detach himself from the common and natural view that Newton's system describes objective truth, but all the same he sticks to his own theory that all systems are the products of human imagination, going beyond the objective evidence and therefore liable to be superseded after a time. It says much for the independence of Smith's mind that he was able to take such a view of Newton's physics in the middle of the eighteenth century.

If Smith was prepared to say that the connecting framework of Newton's astronomical theory was an invention of the imagination, he would *a fortiori* have said it of his own theories of ethics and economics. This does not prevent him from claiming that his own theories have more truth in them than the theories of predecessors, nor does it prevent him from acknowledging, as he does, that Newton's theory of astronomy contains more truth than *its* predecessors. He writes of one alternative theory of economics and one on ethics as approximating to, or bordering upon, the truth, implying that his own theory does better still. An invented theory in science or philosophy implies a number of factual consequences, and if some of these consequences turn out to be false, then the theory can be rejected as false. It may at the same time imply other consequences which correspond with the relevant facts, and on that account it may be described as an approximation to the truth or as bordering on the truth in some respects.

Smith's position is faced with another problem. A philosophical theory about the nature of theories, philosophical and scientific, must refer to itself; and if it says that no theory is a statement of absolute truth, it implies a restriction upon its own truth. Smith's theory that "systems" (scientific or philosophical theories) are inventions of the imagination implies that the theory itself is an invention of the imagination. So why should we believe it? I do not think that Adam Smith would have been greatly worried by this objection. He would have said, as Hume would have said, that the work of the imagination is a natural and inevitable process. No theory can reach absolute truth, but theories can still be rated as more and less true, according to the extent to which they can cover observed facts and can correctly predict observable future facts. Smith would not have been surprised or perturbed that his system of economics has been superseded in many respects.

He would likewise not have been surprised that Newton's theory of physics has been superseded in some respects. Although Newton was the greater thinker, Adam Smith could take a more detached view of scientific (and philosophical) theory, as he could of theology, because he lived a century later and because he could learn from David Hume as well as from Isaac Newton.

NOTES

1 The first reason is elaborated by Kemp Smith; the second by Noxon, building on Hurlbutt.
2 Dugald Stewart, "Account of the Life and Writings of Adam Smith" in Smith, *EPS* 270-1.
3 Adam Smith, "History of Astronomy," IV.74 in Smith, *EPS* 103.
4 Adam Smith, "History of Astronomy," IV.72 in Smith, *EPS* 101.

WORKS CITED

Bentham, Jeremy. *Civil Preface*: Bentham ms. 32. University College, London, qtd. in Halévy.

Buckle, Henry Thomas. *History of Civilization in England*. London, 1857-61. 2 vols.

Halévy, Elie. *La Formation du radicalisme philosophique*. Paris, 1901-04. 2 vols.

Hume, David. *A Treatise of Human Nature*. London, 1739-40. 2 vols.

Hurlbutt, Robert. *Hume, Newton and the Design Argument*. Lincoln: U of Nebraska P, 1965.

Hutcheson, Francis. *An Inquiry concerning the Original of our Ideas of Virtue or Moral Good*. Treatise II of *An Inquiry into the Original of our Ideas of Beauty and Virtue*. London, 1725; qtd. here with modernized spelling, capitalization, and punctuation.

Kemp Smith, Norman. *The Philosophy of David Hume*. London: Macmillan, 1941.

Millar, John. *Historical View of the English Government*. London, 1787; qtd. in *EPS*.

Noxon, James. *Hume's Philosophical Development*. Oxford: Clarendon Press, 1973.

Smith, Adam. *Essays on Philosophical Subjects*. 1795. Ed. W.P.D. Wightman et al. Oxford: Clarendon Press, 1980. Cited as *EPS*.

______*An Inquiry Into the Nature and Causes of the Wealth of Nations*, 1776. Ed. R.H. Campbell and A.S. Skinner. Oxford: Clarendon Press, 1976. Cited as *WN*.

______*Lectures on Rhetoric and Belles Lettres*. Ed. J.C. Bryce. Oxford: Clarendon Press, 1983. Cited as *LRBL*; qtd. here with text amended.

Westfall, Richard S. "Newton and the Scientific Revolution." *Queen's Quarterly* 95 (1988): 4-19

______*Never at Rest: A Biography of Isaac Newton*. Cambridge: Cambridge UP, 1980.

Isaac Newton, Explorer of the Real World

A.P. FRENCH

Introduction

My purpose in this paper is to try to explain, from the standpoint of a working physicist, why I consider Newton to be the greatest scientist who has ever lived. Despite the god-like status that some have tried to ascribe to him, he was of course a fallible human being, and I do not mean merely that he seems to have had a seriously flawed personality. I am thinking much more in terms of his work than of his character. Modern scholarship has amply demonstrated that he groped and fumbled, as all scientists do, in approaching a new problem, that he made mistakes, that he even fudged things a bit sometimes in trying to reconcile theory with observation. But once one has granted all this, I believe that he remains supreme. And at the heart of it, in my opinion, is the unique combination of thinker and doer in one person that he represented.

It is well known that Newton's greatest and most sustained scientific achievements were in mechanics (including the theory of gravitation) and in optics, and I should like to talk in some detail about his explorations in both these fields. But I will also say something about his thoughts and speculations in other areas, because these, too, represent the work of an explorer of the universe around him. My concern will, however, be limited to what we now call physics. I shall be leaving aside not only his protracted labours in chemistry and alchemy, but also his achievements as the foremost pure mathematician of his time.

As has often been said, it would have been difficult to predict Newton's future from his beginnings as a child growing up in the Lincolnshire countryside. He was not an infant prodigy. Nonetheless, there were hints of an inquisitive and probing nature, as for example when he tested the strength of a wind by observing how far he could jump with it or against it, and when, even earlier, he took to marking the movement of shadows on the walls of his house, and so made his own sundials.[1] And he displayed skill and ingenuity in designing and making mechanical models, most notably a toy windmill and a water clock (Christianson 13, 15). But the real Newton emerged, with astonishing suddenness, as

Figure 1. Painting by Sir Peter Lely (1665), said to be a graduation portrait of Newton. [From Newton, Questiones*: frontispiece.]*

he approached the end of his undergraduate career. (Figure 1 is a putative portrait of him at about this time.) The young Newton, still an undergraduate, devoured most of what was then known in mathematics and began making his own original contributions. He also began his investigations into optics, and it is with these that I shall first concern myself.

Newton's work in optics is dominated by a succession of controlled experiments, out of which he built up a picture of what light is and how it behaves. His first concern seems to have been with the way in which the eye perceives colour. The eye as an image-forming device was already well understood, but colour perception was much more mysterious. (It still is.) One of the twenty-one-year-old Newton's first experiments, in 1664, took the form of pressing one side of his eyeball with a fingertip and observing the coloured rings that appeared to be formed around the pressure point (Newton, *Questiones* 438). A little later he did the truly horrifying experiment of thrusting a bodkin into his eye socket, between the bone and the eyeball, to study the same phenomenon closer to the centre of the retina (Fig. 2). He also stared directly at the sun and studied the subsequent colour impressions he obtained by looking at light or dark objects. It is a miracle that he did not blind himself or do himself serious injury.

Very soon, however, he concentrated his energies on investigating more objectively the refraction of light through prisms. He found that if a strip of paper was viewed through a prism after being painted half red and half blue, the blue half appeared to be more laterally displaced than the red half (Fig. 3). Then he began a systematic investigation of the white light from the sun. Here his

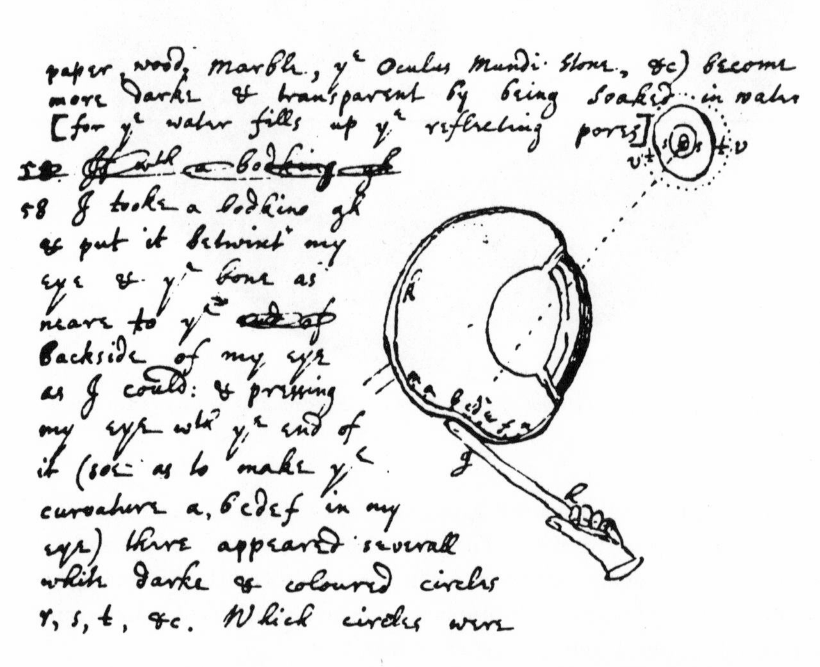

Figure 2. Newton's experiment on colours produced by pressing his eyeball. [From Westfall 95.]

Figure 3. *Newton's experiment on the splitting of red and blue images as seen through a prism. [From Westfall 165.]*

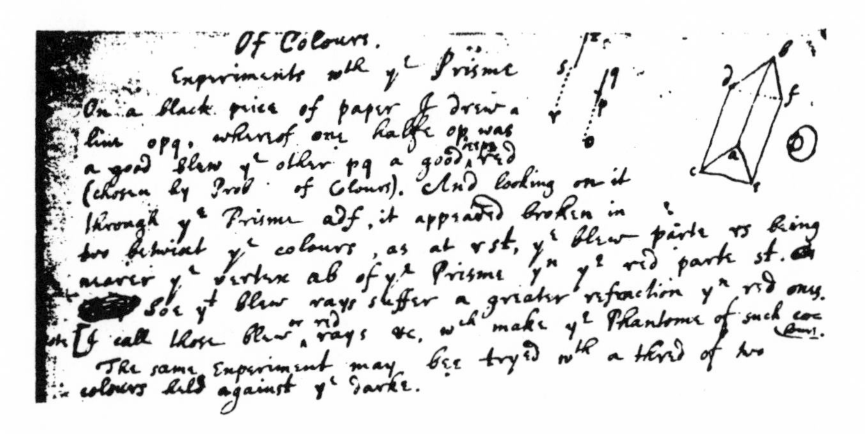
Of Colours.
Experiments wth ye Prisme
On a black peice of paper I drew a line opq, whereof one halfe op was a good blew ye other pq a good deepe red (chosen by Prob. of Colours). And looking on it through ye Prisme adf, it appeared broken in two twixt ye colours, as at rst, ye blew parte rs being nearer ye vertex ab of ye Prisme yn ye red parte st. Soe yt blew rays suffer a greater refraction yn red ones.
[I call those blew or red rays &c, wch make ye Phantome of such colours.]
The same Experiment may bee tryed wth a third of two colours laid against ye darke.

full powers as an experimentalist revealed themselves – the brilliantly analytical mind behind the physical arrangements that it devised. The fact that sources of white light appeared to have brightly coloured edges when viewed through a prism was already familiar, but Newton was the first person to demonstrate the underlying basis of the phenomenon. He did this, in the first instance, by letting a narrow beam of sunlight fall on a prism in a darkened room, and then allowing the refracted light to travel a considerable distance (about twenty feet) before striking a screen.[2] The arrangement is quaintly but nicely illustrated in an English translation of Voltaire's *Eléments de la Philosophie de Newton* (Fig. 4).[3]

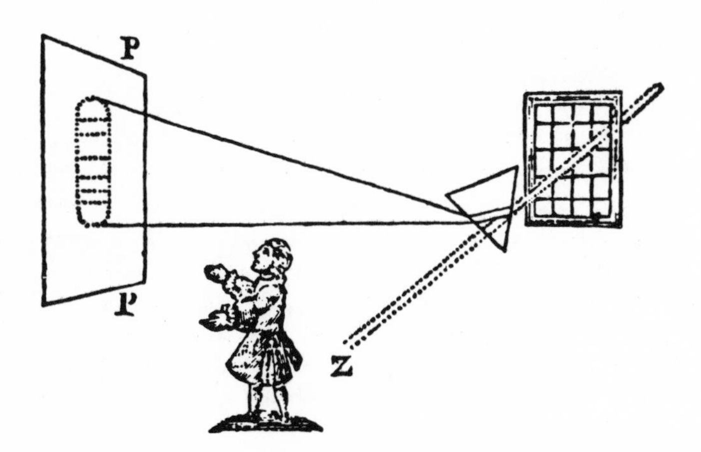

Figure 4. *One of Newton's first experiments in forming an extended spectrum. [From Voltaire 97.]*

It had been the belief that colours were the result of the modification of pure white light by materials. What Newton concluded was that the various colours were already present in the white light, that it was itself complex, and the prism merely resolved it into its various components. He recognized that the elongated strip of colour, red at one end and violet at the other, was in fact a succession of partially superposed images of the sun formed by the prism. The width of the strip was exactly what could be expected from the size of the hole and the angular diameter of the sun. The length of the strip, in his first experiments, was about five times this. In later experiments, using first a collimating aperture close to the prism and later a converging lens, he was able to greatly reduce the diameter of the sun's image (Fig. 5). Under these conditions he finally obtained a spectrum forty or more times longer than it was wide.

Newton chose to describe the spectrum as being a resolution of white light into seven colours – the traditional sequence (as it subsequently became) of red, orange, yellow, green, blue, indigo, violet. This number of colours was in deliberate analogy to the seven notes of a musical scale, up to but not including the octave. (More about this analogy later.) But Newton understood that in fact there was no natural division into discrete parts; the spectrum was continuous and with better resolution the variety of distinguishable colours became progressively greater.

These first observations on the resolution of white light into a spectrum were

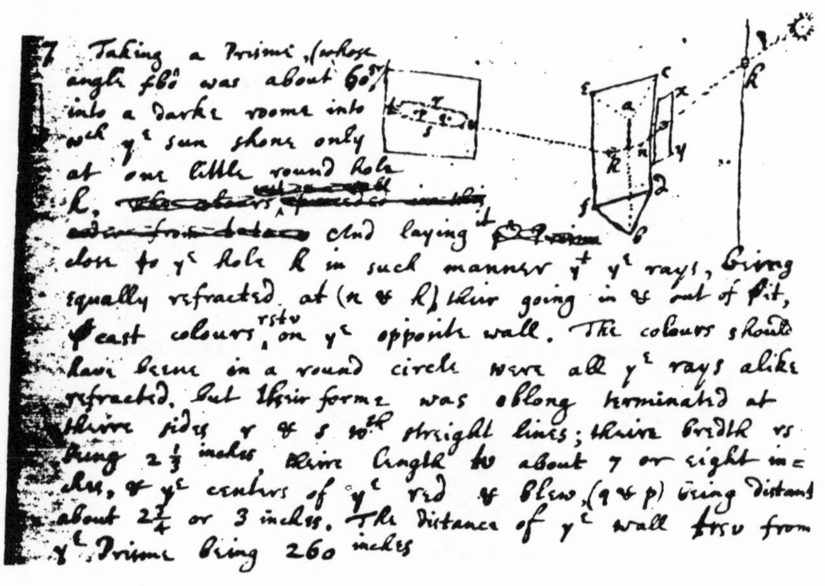

7 Taking a Prisme (whose angle fbd was about 60gr) into a darke roome into wch ye sun shone only at one little round hole k, And laying it close to ye hole k in such manner yt ye rays, being equally refracted at (n & h) their going in & out of it, cast colours rstv on ye opposite wall. The colours should have beene in a round circle were all ye rays alike refracted, but their forme was oblong terminated at their sides r & s wth streight lines; their bredth rs being 2 1/3 inches, their length tv about 7 or eight inches, & ye centers of ye red & blew (q & p) being distant about 2 3/4 or 3 inches. The distance of ye wall trsv from ye Prisme being 260 inches

Figure 5. An improved version of Newton's first experiment on the formation of a spectrum. [From Westfall 165.]

Figure 6. Newton's crossed-prism experiment. [From Newton, Opticks *36.]*

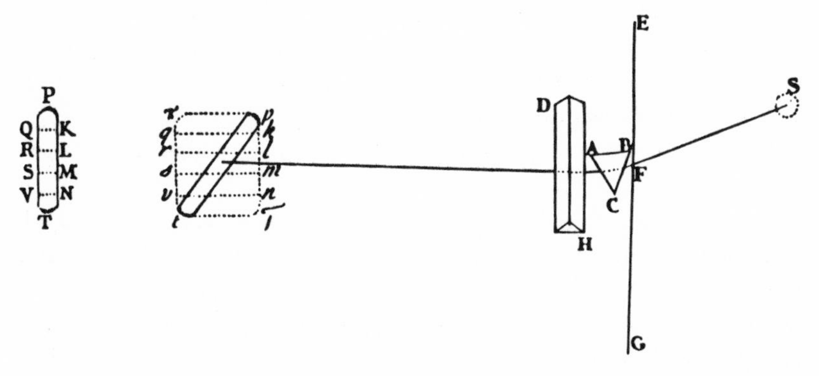

by no means enough to satisfy Newton. In all kinds of ways he sought to demonstrate the correctness of his analysis. One very beautiful experiment was to pass the light emerging from a first prism through a second prism oriented at right-angles to it (Fig. 6). This showed that no further modification of the light occurred. Once the separation of white light into its various components had been achieved with the first prism, the second prism simply refracted the individual colours by differing amounts and produced a final spectrum along a diagonal line, rather than a pattern filling a rectangle.

To show that differing refractions for differing colours were the fundamental process, Newton devised what he later called the crucial experiment (*experimentum crucis*) in which, with the help of two prisms and two auxiliary apertures (Fig. 7), he was able to demonstrate that light of different colours, falling at the *same* angle on the second prism, underwent different angles of devia-

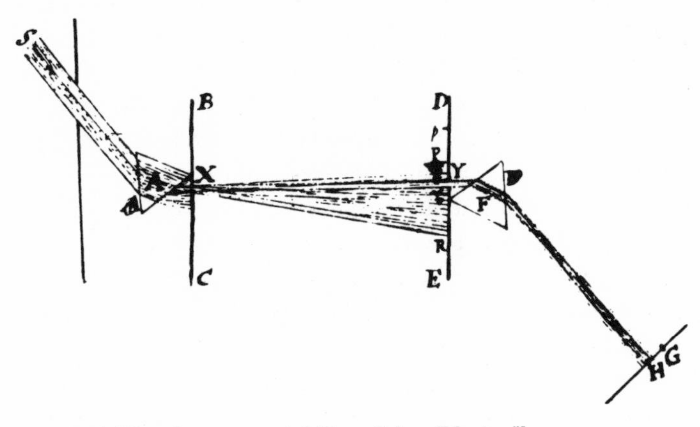

Figure 7. Newton's "Experimentum crucis." [From Cohen, "Newton."]

tion there. (By turning the first prism he could select light of any chosen colour to pass through the hole X and arrive at Y.)

To confirm that the phenomena were a property of the light and not of coloured objects, he went back to his original experiment with the red/blue strip of paper, and showed that the same splitting of images seen through the prism could be obtained if one viewed, in a similar way, a pure white thread, one half of which was illuminated with red light and the other half with blue (Newton, *Opticks* 49).

What one sees here (and I have described only a fraction of his investigations) is something for which I find no precedent or parallel in the scientific literature of Newton's day. Not that there was any shortage of experimentation. The *Philosophical Transactions* of the Royal Society of London (which had been founded in 1662) are full of accounts of intriguing observations. But there was absolutely nothing to compare to Newton's sustained and many-faceted attack on a single problem. Even in our own day, his approach could be taken as a model for the exploration of a new phenomenon. It was something qualitatively different from the mere collection of precise observations which in astronomy, at any rate, had been a very highly developed art for thousands of years.

In 1669, at the age of twenty-six, Newton was appointed Lucasian Professor at Cambridge University. Although this chair was in mathematics, Newton chose to give his inaugural course of lectures on the subject of optics and his own optical researches. Their original title was *Lectiones Opticae* (the text was in Latin, and so, presumably, was the oral presentation of the material). In a revised form, and under the title *Optica*, these lectures were deposited with the Cambridge University Library, as required by statute, perhaps as late as 1674.[4] Their final section – another Newtonian *tour de force* – was a quantitative explanation of the rainbow. The groundwork had already been laid by Descartes, and before him by Antonio de Dominis, Archbishop of Spalato. It was realized that the basis of the rainbow was the refraction and internal reflection of the sun's rays by individual water droplets (Fig. 8). Descartes had even understood that the angle subtended by the bow at the eye corresponded to a preferred direction of emergence of the light after its passage around the inside of the drop (the phenomenon of minimum deviation). But Newton, with his knowledge of the exact refractive power of water for different colours, was able to go on to show explicitly how the primary rainbow should lie between 41° (blue) and 43° (red), and the secondary bow between about 49° (red) and 53° (blue). Figure 9 (from the *Opticks*, the much later English-language account of his optical researches) shows a clear picture of the formation of these bows, as good as any that one would find today. It was a beautiful piece of work, but one must acknowledge that this kind of beauty is not universally appreciated. There is

Figure 8. Refraction and reflection of light within a water drop. [From Newton, Opticks *170.]*

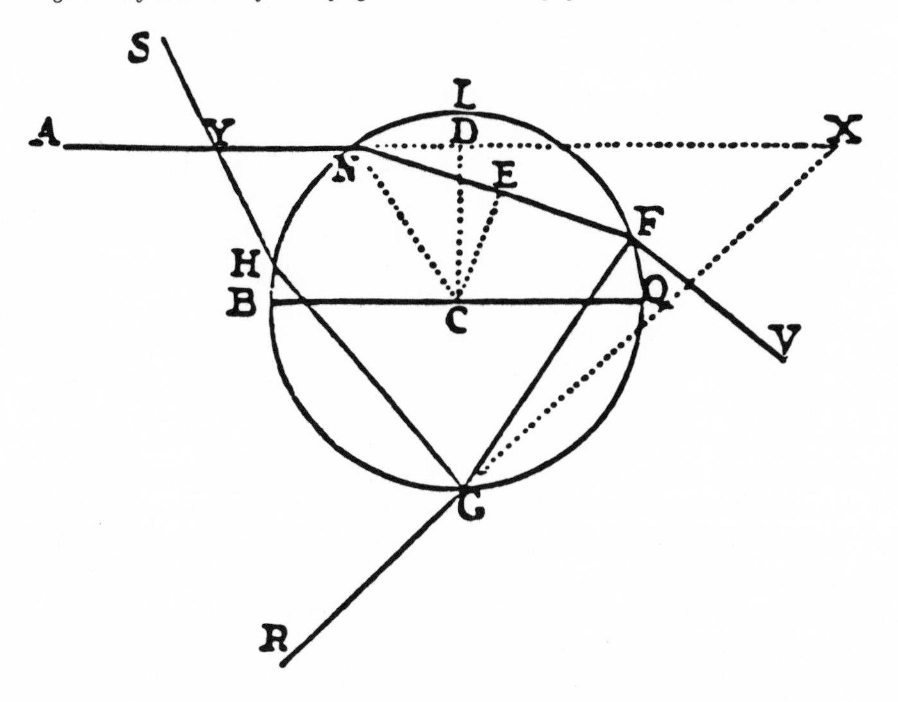

Figure 9. Formation of primary and secondary rainbows. [From Newton, Opticks *173.]*

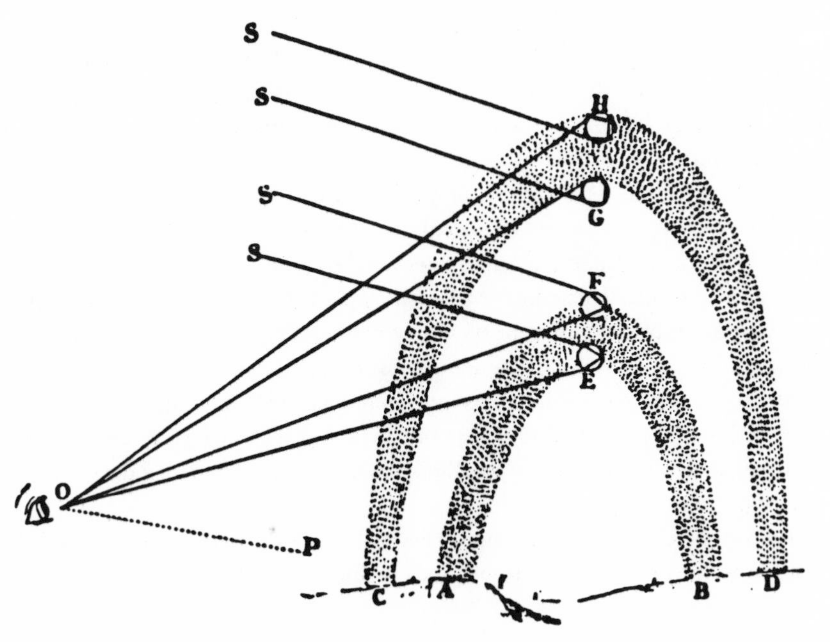

a story of a rather drunken dinner party given by the English painter Benjamin Haydon in 1817. Keats, Wordsworth, and Charles Lamb were there. Keats reportedly proposed a toast to the confusion of Newton, and agreed with Lamb that Newton "had destroyed all the Poetry of the rainbow by reducing it to a prism" (Gjertsen 268).

One footnote to this account of Newton's study of the spectrum – small in itself but indicative of his amazing skill as well as his searching spirit – is that in one experiment he used Venus instead of the sun as the source of light, and was pleased to see that its refracted image beyond the prism was "drawn out into a long splendid line" (Newton, "Oldenburg" 13.4.72).

As a result of these researches on the separation of colours by refraction, Newton reached the conclusion that there were fundamental limits on the sharpness of the image obtainable with a glass lens. He showed by direct experiment that the focal length of a given lens was significantly different for red light and blue light (Newton, *Opticks* 86). (The human eye, of course, shares this property, which accounts for the acute discomfort you may have felt when looking at, say, a poster in which bright blue and bright red areas of colour are juxtaposed.) Thus a white point-like object, such as a star, cannot possibly form a correspondingly point-like image. In addition to this phenomenon of chromatic aberration there is another effect – spherical aberration – that impairs the focal quality, but even in Newton's day the makers of telescopes knew that this should be curable by grinding the surfaces of lenses to contours slightly different from spherical. As a practical matter, however, the chromatic aberration could not be eradicated, although Newton did apparently recognize the theoretical possibility of making achromatic combinations using different types of glass. Faced with this problem, Newton proceeded to design and build the first reflecting telescope. It was a remarkable achievement, which he describes in detail in the *Opticks* (102-111). The mirror, ground and polished by himself, was made of a copper-tin-arsenic alloy (speculum metal) that he had prepared according to his own recipe; he also made all the other parts. A design drawing that he made (Fig. 10) demonstrates his thorough mastery of mechanical design. The result of his efforts was a telescope, with a tube only about eight inches long, that magnified nearly forty diameters. Newton was justifiably proud of it. He wrote to an unidentified friend: "I have seen with it Jupiter distinctly round and his Satellites, and Venus horned" (23.2.68/69). He also reported to Henry Oldenburg, Secretary of the Royal Society, that he could distinguish words on a page of the *Philosophical Transactions* at a distance of well over 100 feet (19.3.1671/72).

It was this telescope that brought Newton's extraordinary talents to the attention of the officers of the Royal Society and, through them, to the rest of

Figure 10. Newton's own design drawing for one of his reflecting telescopes. [From Newton, Correspondence *1:Facing 76.*]

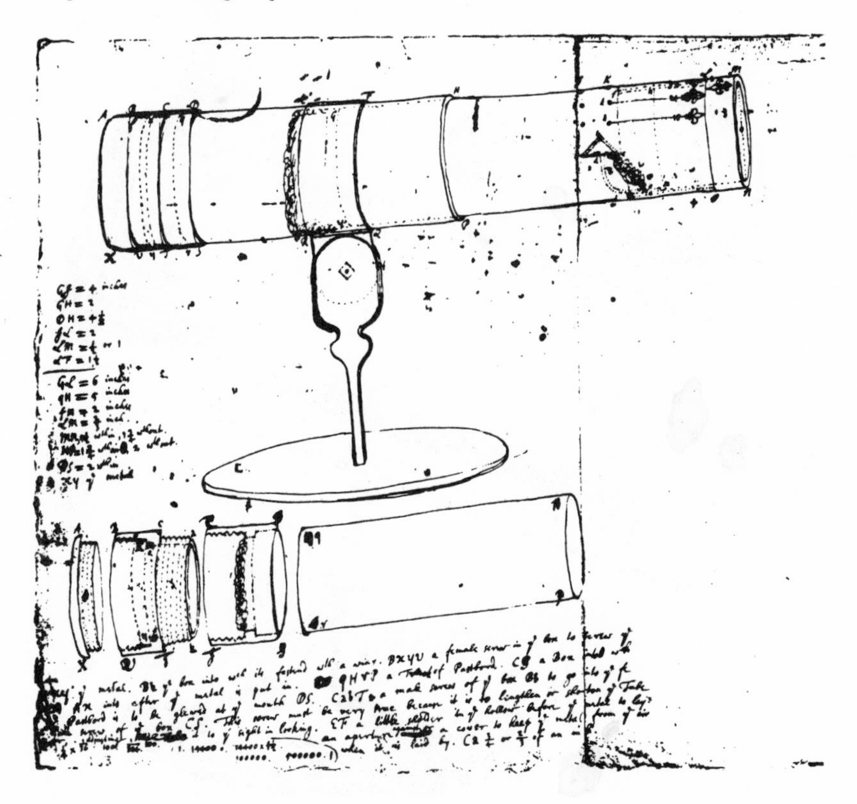

Europe. They asked to see the telescope, and in late 1671 a second version of it was delivered to London. Figure 11 shows sketches of an ornament on a weathervane three hundred feet away, as seen (Fig. 2) through Newton's telescope and (Fig. 3) through a refracting telescope twenty-five inches long. (The sketches were made, much as a biologist may sketch what he sees through a microscope, by using one eye to look into the instrument as the other eye guides the hand in copying the image, with its perceived size, onto a piece of paper about a foot from the eye.) Within about a month, in January 1672, Newton was elected a Fellow of the Royal Society, and that same year saw the publication in the *Philosophical Transactions* of no less than eight communications from Newton regarding his work in optics (Cohen, *Newton's Papers*).

Before leaving the subject of Newton and optics, I need to mention one other very important area. Newton's optical lectures were concerned only with the reflection and refraction of light (including of course the resolution and com-

Figure 11. Newton's reflecting telescope, with sketches of images of a distant object seen through it (Fig. 2) and through a refracting telescope four times longer (Fig. 3).

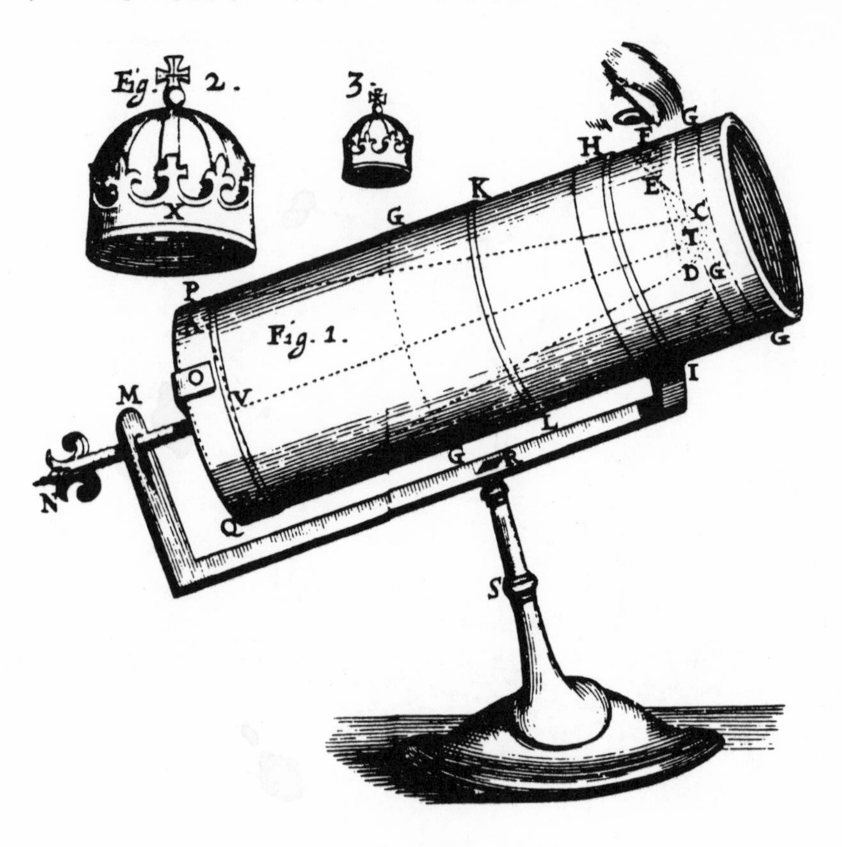

position of colours) but he had also, by about 1670, embarked on a study of the phenomena observable with thin films (such as soap bubbles). In typical fashion he defined and quantified the situation by forming an air gap of varying thickness between two glass surfaces, one flat and the other very gently convex, and observing the bright and dark rings seen when one looked down from above (Fig. 12). He measured the radii of the rings to an estimated accuracy of better than one-hundredth of an inch, and discovered that the squares of these radii were in arithmetic progression. This meant that the successive rings corresponded to equal increments in the thickness of the gap between the glass surfaces. He went further, and measured the characteristic changes of thickness for different colours. As in his researches on the spectrum, he presented the results in terms of the visual equivalent of a musical octave. What he was seeing has come to be fully explained in terms of a wave theory of light. But Newton had become committed to a corpuscular theory, for reasons I will not go into

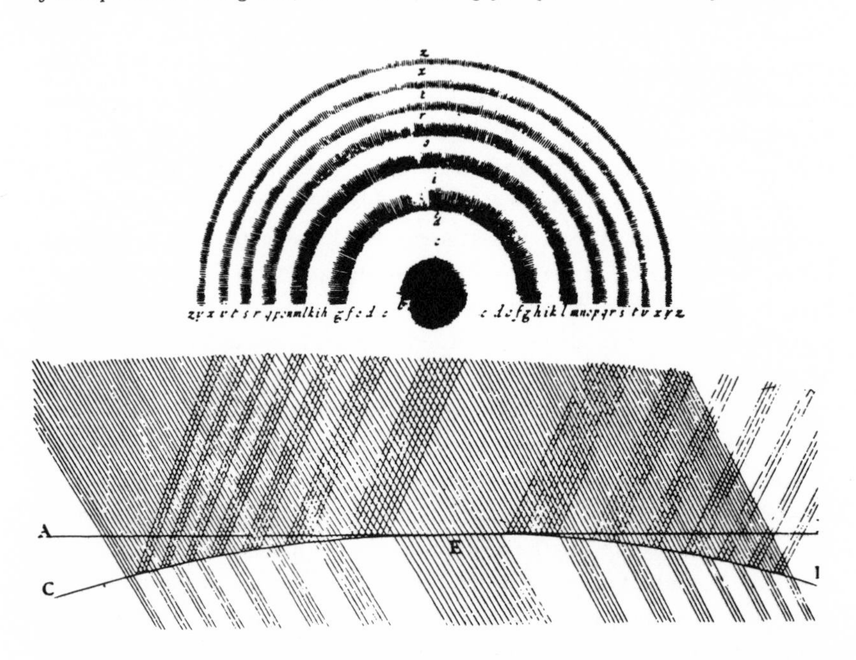

Figure 12. "Newton's rings" (he was not their first discoverer). The lower part of the figure, a side view of the experimental arrangement, shows the alternating "fits" of transmission and reflection.

just here. He interpreted the bright and dark rings as being due to what he called "fits" of reflection or easy transmission for the particles of light encountering the gap (*Opticks* 281-82). He thought that somehow these particles set up vibrations, like pebbles dropped into a pond, in some subtle medium (the "ether") that pervaded all space, and that this decided the outcome. If only he could have seen things differently! For in measuring the change of air gap from one "fit" to the next, he had in fact determined, with excellent accuracy, the value of half a wavelength for different colours. And this change of wavelength, as one goes from blue to red, is almost a factor 2. The physical correspondence with the musical octave happens to be amazingly close. But Newton did not accept this as definite evidence for the wave nature of light itself. We should not, however, be too critical, for we have learned in the twentieth century that both particle and wave properties are needed to describe the properties of light.

The Development of Newton's Mechanics

Let me now turn to Newton's mechanics, which paved the way for his most glorious achievements. As is well known, Newton wrote only two books. His *Opticks*, although not published until 1704, was based on the work I have been describing, which was completed in the 1670s. After that time Newton went on to other things. If you have looked at the *Opticks*, you will know that it is written in an easy conversational style. Newton gave a vivid and detailed narrative account of his experiments and the conclusions he drew from them. But his great masterpiece, the *Principia*, presents a very different aspect. Quite apart from the fact that its first editions were in classical Latin, the whole presentation is forbidding. Using Euclidean geometry with a virtuosity that few mathematicians today could emulate, Newton presented his "Mathematical Principles of Natural Philosophy" in a Euclidean framework of axioms, lemmas, and corollaries. At least, this is the general character of the first two books. In the third book, *The System of the World*, Newton proceeds to apply his theory to all the celestial phenomena that were known or knowable in his time. It was a dazzling performance. Behind it, however (or perhaps one should say before it), was a wonderful struggle on Newton's part to understand the bases of a mechanical universe. For the *Principia* was not published until 1687, but already in 1666, more than twenty years earlier, the young Newton had begun to speculate on the motions of the moon and the planets. There seems to be no reason to doubt the basic truth of the story of Newton and the apple – how, in 1666, having left Cambridge for a while on account of the Great Plague, he was moved by the fall of an apple in his Lincolnshire garden to speculate if the moon, too, was falling towards the Earth in a similar way (Herivel 65-69). He was able to calculate that in one second, while travelling about one kilometre in its orbit, the moon deviates from a straight-line path by about a twentieth of an inch. In the same length of time an object projected horizontally on the Earth would fall about sixteen feet. The ratio, about 3700, was very close to the square of the ratio of the moon's distance to the Earth's radius. Was this significant? Newton was in no position to answer that question at the time. But he was also aware of Kepler's discovery that the squares of the periodic times of the planets are in proportion to the cubes of their mean distances from the sun. He deduced, to quote his own words (written before 1669) that "the endeavour of receding from the Sun will be reciprocally as the squares of the distances from the Sun" – that is, an inverse-square law of centrifugal force (Westfall 152). The germ of Newton's theory of universal gravitation was undoubtedly present here, but his full understanding and exploitation of the theory did not take place until nearly twenty years later, when he returned to mechanics in earnest. The culmination of this effort was the *Principia*, the first edition of which was published in 1687. Figure 13 shows Newton at about this time – actually in 1689, when at the age of 46 he began a brief tenure as Member of Parliament for Cambridge University. Figure 14, from a contemporary print of Trinity College, shows where Newton lived and worked during this peak period of his career – his rooms on the first floor beside the main gate, and below them the garden and (against the chapel wall) the laboratory in which he conducted many of his experiments, especially in chemistry and alchemy.

It is not my intention here to discuss the *Principia* as such; my focus will continue to be on Newton, the explorer of nature at first hand. But I do need to say something of the context in which he presented his specific discoveries. Book I, *The Motion of Bodies*, is a development of the basic dynamics of particles, and embodies the theoretical substructure for the celestial mechanics of Book III. Book II, *The Motion of Bodies in Resisting Mediums*, had as its main result, and

Figure 13. Portrait of Newton by Godfrey Kneller. [From Newton, Correspondence *I: frontispiece.]*

Figure 14. Portion of David Loggan's print of Trinity College, Cambridge, from Cantabrigia Illustrata *(1690). Newton's rooms, garden, and laboratory, between the main gate (left) and chapel (right). [From Royal Society* Celebrations *Facing 30.*]

its main goal, a demonstration of the invalidity of Descartes's hypothesis that the circular motions of heavenly bodies resulted from their being swept around in huge fluid vortices (Descartes, *Principia* 8: 80-202). So Book I is concerned with the general laws of motion, and Book II with fluid mechanics.

In Book I we see Newton first presenting his ideas about space and time. Almost at once we encounter a real experiment. Newton believed in an absolute space. But how could we know anything about it? Observed positions and velocities are always relative. But Newton believed that acceleration – the central quantity in his laws of motion – was absolute. The proof? Take a bucket of water hanging at the end of a tightly twisted rope, and let it spin (*Principia* 10). At first the bucket turns but the water remains at rest and its surface remains flat. But then the water picks up the rotation, and its surface becomes curved. Clearly it was not nearby objects that defined the frame of reference for the centripetal acceleration associated with this rotation, but something more fundamental, closely related to the space defined by the fixed stars. The deep implications of this experiment have continued to be debated into our own day.[5]

On a less abstract level, Newton's laws of motion were based on some very simple experiments: observations of collisions between different kinds and sizes

Figure 15. Pendulum experiment, from Principia, Book *1. [From* Motion of Bodies *22.]*

24 PHILOSOPHIÆ NATURALIS

AXIOMATA, SIVE

decim & reſtabit nihil: ſubducantur aliæ partes duæ, & fiet motus duarum partium in plagam contrariam: & ſic de motu corporis *B* partium ſex ſubducendo partes quatuordecim, fient partes octo in plagam contrariam. Quod ſi corpora ibant ad eandem plagam, *A* velocius
5 cum partibus quatuordecim, & *B* tardius cum partibus quinque, & poſt reflexionem pergebat *A* cum quinque partibus; pergebat *B* cum quatuordecim, facta tranſlatione partium novem de *A* in *B*.
10 Et ſic in reliquis. A congreſſu & colliſione corporum nunquam mutabatur quantitas motus, quæ ex ſumma motuum conſpirantium & differentia contrariorum col-
15 ligebatur. Nam errorem digiti unius & alterius in menſuris tribuerim difficultati peragendi ſingula ſatis accurate. Difficile erat, tum pendula ſimul demittere ſic, ut corpora in ſe mutuo impingerent in loco infimo *AB*; tum loca *s*, *k* notare, ad quæ corpora aſcendebant poſt concurſum. Sed & in ipſis corporibus pendulis inæqualis par-

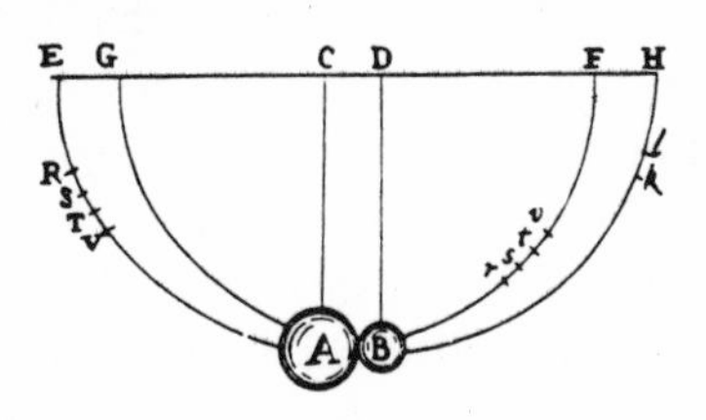

of objects. From them Newton extracted his quantitative concept of momentum and the role of force in changing the momentum of a given body. The groundwork had been laid in 1668 when Christopher Wren, John Wallis, and Christiaan Huygens, at the invitation of the Royal Society, studied these phenomena. The experiments were made possible by a wonderful property of the simple pendulum: no matter from what angle it is released, a pendulum takes almost exactly the same time to reach its lowest point. Thus, using two pendulums of the same length, one can produce collisions at the bottom point for a variety of conditions. At the instant of collision, the two bodies are travelling horizontally; afterwards they recoil to positions that depend on their velocities immediately after the collision. Figure 15 illustrates the arrangement. Newton, in his masterly way, made corrections for the losses due to friction, and demonstrated that the conservation of momentum is a fundamental law of physics.

In another pendulum experiment, also described in the *Principia*, Newton demonstrated one of the most important and fundamental relationships in physics, the proportionality of weight to mass. He did this by showing that pendulums of the same length all had the same period of swing, regardless of the mass or composition of the pendulum bob (*System* 411). This implied (using his laws of motion) that the gravitational force acting on any object due to the Earth is proportional to its mass. It was a crucial step towards the theory of universal gravitation. And Newton, even though he did not have access to sophisticated timing devices, demonstrated this proportionality to better than one part in a thousand.

The pendulum played an important role in many other of Newton's researches. One of the most fundamental questions at the time (and until the beginning of the present century) concerned the existence and possible properties of the "ether" that Newton himself had invoked in trying to explain the colours of thin films. It was imagined to permeate space and to penetrate all material objects. But in moving through this medium, should not objects encounter resistance? Newton tested this by making a pendulum eleven feet long and studying the decay of its oscillations when the pendulum bob was an empty box, or the same box filled with dense metals (*Resisting Mediums* 325-26). His conclusion was that no discernible effect of ethereal resistance was present.

Again, using the pendulum as a clock, Newton investigated the laws of resistance to motion through real fluids. (This was in Book II of the *Principia.*) He learned that the main effect is a resistive force proportional to the square of the speed of the moving object and to its area of cross section, a basic result in hydrodynamics ever since (328-31). The thirteenth and last experiment that he cites on this matter involves an interesting bit of history. It was done in 1710 by Francis Hauksbee (experimental assistant at the Royal Society), no doubt at Newton's instigation (Hauksbee). The main fabric of the new St Paul's Cathedral, built by Wren after the Great Fire of London, had just been completed, and Hauksbee compared the times of fall of light and heavy spheres dropped from the top of the cupola to the main floor, 220 feet below.

Finally, as one more by-product of his studies of fluid motion, Newton developed a theory of the speed of waves in fluids, and tested it by measuring the speed of sound in air (*Resisting Mediums* 378-84; Westfall 456, 734-36). Using the newly erected Nevile's Court at Trinity College as his laboratory, he adjusted a short pendulum until he judged that its period coincided with the interval between echoes of sound passing back and forth between the end walls of the court. He found a speed of about 1000 feet per second. This was one instance where Newton tinkered with his theory. His calculated speed was about ten per cent lower than the observed. He then postulated that this was due to the finite size of the particles of air, what he called their "crassitude" (i.e., coarseness). He estimated (quite correctly) that in air the diameters of the particles ("molecules" to us) take up about a tenth of their separation. This, he decided, reduced by the desired ten per cent the time needed for the sound to propagate from one particle to the next. In fact the true speed of sound is about twenty per cent bigger than Newton's original calculation would have predicted, and the reason for the discrepancy is quite different from what Newton proposed. But then theorists everywhere have always been much too ready to trim their theories to fit the currently accepted experimental results!

The Mechanics of the Universe

The work that I have described so far would already have been more than enough to mark Newton as one of the greatest scientists of all time. But for me, and I think for most people, including his contemporaries, the crowning glory of his work was his *System of the World* (*Principia*, Book III) in which he analyzed the workings of the universe as a magnificent machine.

It is hard to know where to begin, but perhaps as good as anything is where Newton himself began, with the satellites of Jupiter (Fig. 16). The discovery of these satellites by Galileo in 1610, using one of the earliest telescopes, was, of course, one of the great events in astronomical history – the first additions to the solar system since antiquity. Hardly less remarkable was the way in which Galileo himself immediately proceeded to record their positions, night after

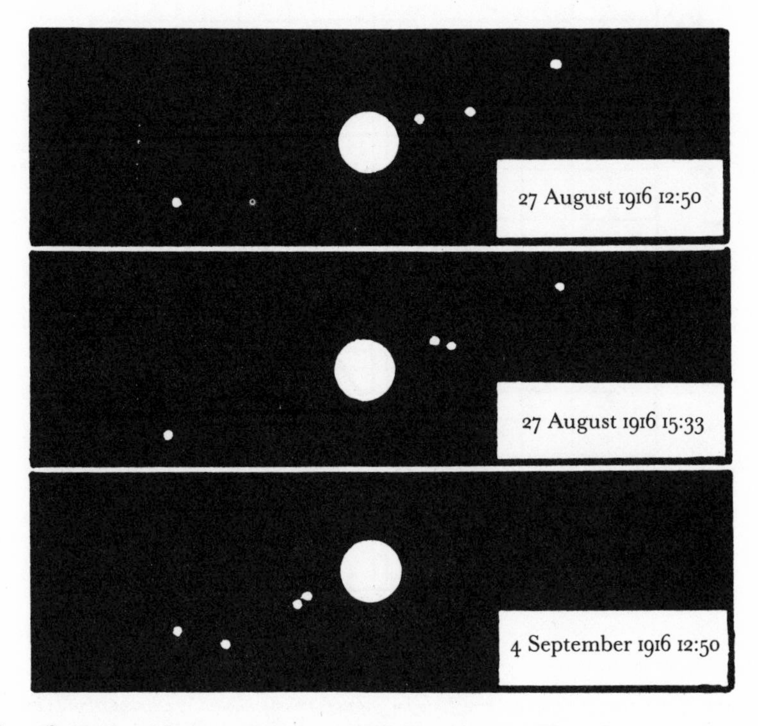

Figure 16. Jupiter and the four moons discovered by Galileo. After a set of Yerkes Observatory photographs taken at somewhat different times.

night, for many months. As a result he was able to state their orbital periods with considerable accuracy. The data showed clearly that Jupiter's satellites, like the planets going around the sun, conformed to Kepler's Third Law: the square of the period of each planet (the time for completing its journey around the sun) is proportional to the cube of its mean distance from the sun (Fig. 17). For Newton this was powerful evidence that a truly universal inverse-square law of gravitation was at work. He asked John Flamsteed, the first Astronomer Royal, to provide up-to-date observations. In fact, as Figure 17 shows, Flamsteed's results were not greatly different from Galileo's. And then, in 1687 (too late for the first edition of the *Principia*, but cited in the second edition) the astronomer Giovanni Cassini reported corresponding data for the five recently discovered satellites of Saturn (four of them credited to himself). From all this information, plus his fundamental law of gravitation, Newton was able to compare the masses of Earth, sun, Jupiter, and Saturn (*System* 416), because the motion of any satellite involves a force proportional to the central mass that

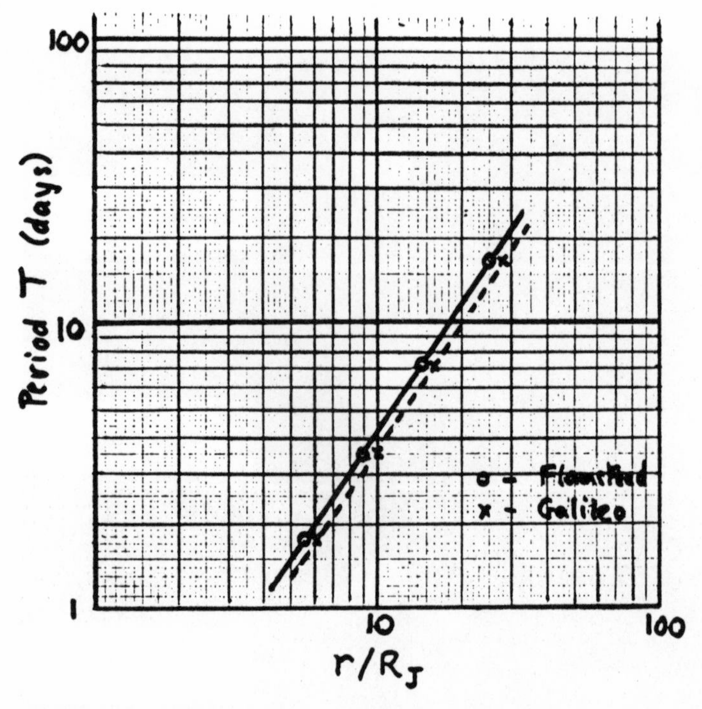

Figure 17. Verification of Kepler's Third Law for Jupiter's moons, shown by a linear relation between the logarithms of the periods and the logarithms of the orbit radii, with a slope of 3/2. (The orbit radii are expressed as multiples of the radius of Jupiter.)

attracts it, and this fact is reflected in the orbital period. Newton, armed with his theories, had thus gained access to hitherto unknowable facts about the universe. This was exploration on a grand scale.

Newton, however, was also alive to the more local implications of his gravitational law. Figure 18, a famous illustration from a popularized version of the *System of the World*, shows his pictorial way of demonstrating how there is a uniform progression from the path of a projectile, launched horizontally above the Earth's surface, to the possibility of satellites in elliptical orbits around the Earth at different distances from the centre. The parabola followed by a projectile near the Earth is in fact more accurately described as a small part of an ellipse, interrupted where the orbit intercepts the Earth's surface.

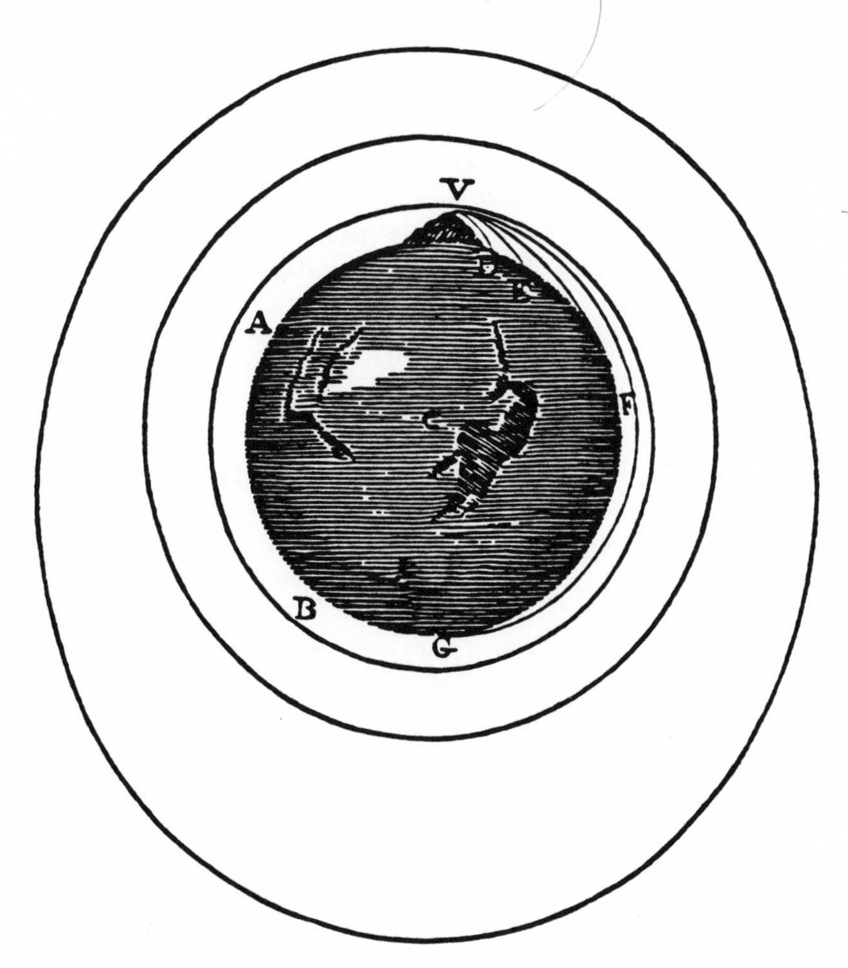

Figure 18. Newton's view of possible earth satellite orbits. [From Newton, System *551.]*

Another of Newton's great achievements was his application of universal gravitation to the motions of comets. Against all the folklore that had surrounded comets through the ages, he placed the rigour of careful observation and mathematical analysis. It was not an easy task. To establish the precise shape of a comet's trajectory on the basis of observations from the moving Earth was in fact incredibly difficult. Newton had first undertaken detailed observations of a comet in 1664, while still an undergraduate (*Questiones* 298-302, 357-59, 412-16), but his interest in the matter subsided until December 1680, when a spectacular comet became visible, moving away from the sun towards the outer reaches of the solar system. Newton himself charted its course night after night. At the same time John Flamsteed was following it from the Royal Observatory at Greenwich. And Flamsteed was familiar, also, with a less brilliant comet that had been seen approaching the sun only a month earlier. He proposed to Newton that these were one and the same object. Newton at first strongly resisted this suggestion; such a complete reversal of direction seemed inexplicable to him at the time. But the idea grew on him, and when, in the process of writing the *Principia* in 1685, he made a full assault on the problem of comets, he took the comet of 1680/81 as his chief case (*System* 504-21). (In the meantime, in 1682, his friend Edmund Halley had observed the comet that came to bear his name.)

Comets, in general, are such small objects that they are invisible until they enter the inner regions of the solar system. And here their trajectories can be excellently fitted as parabolas. In fact the orbit of a comet is either hyperbolic, in which case it makes only one close encounter with the sun, or a very elongated ellipse, in which case it returns periodically (like Halley's comet, with its period of 76 years). Halley believed that the comet of 1680 was to be identified with a comet said by Plutarch to have appeared in the year of Julius Caesar's death (44 BC) and subsequently made famous by Shakespeare:

> When beggars die there are no comets seen;
> The heavens themselves blaze forth the death of princes.
>
> *Julius Caesar* (II.ii.30)

Halley also thought that this comet had made two other recorded appearances in between, giving it a period of 575 years. Newton's theory of orbits would then have implied that it was travelling in an enormously elongated ellipse, almost grazing the sun at perihelion and then going out as far as 140 times the Earth's mean distance from the sun. Newton duly reported this in the *Principia*. Later opinion, however, has not upheld this rather romantic view. It is now believed that the comet of 1680 was a once-only visitor. But that change does not in any way impair the validity of Newton's analysis of the comet's trajectory over its visible path.

Newton did not rest content with analyzing the comet's orbit. He proceeded to speculate on the actual composition of such an object. He estimated that the heating by the sun, as the comet's head passed at a distance of only about half a million miles from the sun's centre, was more than enough to vaporize any solid substance (*System* 521-25). He deduced that the comet's tail was made of such vaporized material, which explained why these tails were much bigger and brighter after the perihelion point than before. Figure 19 is Newton's own careful drawing of the 1680 comet's orbit and the appearance of its tail almost to the orbit of Mars, showing clearly how the tail was always directed away from the sun – blown, as we now know, by solar radiations of various kinds.

Two other of Newton's great calculations involved the Earth itself. The first was his explanation of the tides (*System* 435-40). Using once again the inverse-square law of gravitational attraction, he showed how the moon and the sun would combine to create a double bulge in the ocean, with the maxima diametrically opposite one another. A given place on the spinning Earth would then encounter two maxima and two minima of ocean level every day. The actual behaviour of the tides is much more complicated than this, as Newton recognized, but he did understand clearly the way in which the effects of sun and moon combine differently at different times to produce the monthly cycle of tides of different heights.

Another much more subtle phenomenon, but which Newton recognized as being due to the same basic cause, was the very gradual shift in the direction of the Earth's axis of rotation: although today it points towards the Pole Star,

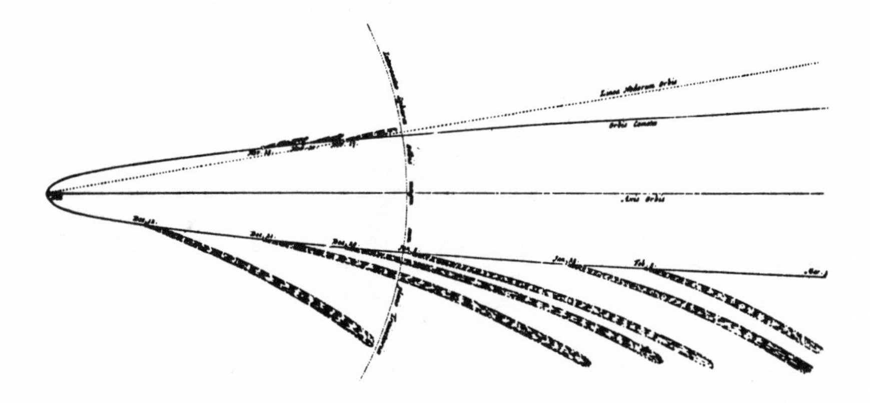

Figure 19. Orbit of the comet of 1680/81. Drawing by Newton. From the first edition of Principia *[From Westfall 463.]*

observations by ancient astronomers (and first correctly interpreted by Hipparchus in the second century BC) showed that this had not always been the case, and that the celestial pole is in fact tracing a circular path among the stars, taking about 25,000 years to complete one circuit (Berry, *Short History* 51). Newton was the first person able to supply a physical explanation for the phenomenon. He had already estimated that the Earth, by virtue of its spin, must have an equatorial bulge equal to about 1/300 of its radius (*System* 424-27). He then considered the way in which the moon and the sun would pull on this bulge (*System* 489-91). Since the polar shift is so very slow compared to a month or a year, one can imagine the material of the moon (and likewise of the sun) to be in effect smeared out into a thin circular ring. Such a ring, exerting its gravitational forces on the Earth's equatorial bulge (Fig. 20), tends to twist the Earth's axis into a direction perpendicular to the plane of the moon's (or the sun's) orbit. But the spinning Earth is like a gigantic gyroscope, and its response to this torque is not a realignment, but a precession, in which its spin axis traces out a conical path from east to west. This was what the ancient astronomers had discovered as the precession of the equinoxes. Newton did his best to obtain a rate of precession in exact agreement with observation. His calculation cannot be defended in detail, but in principle, and in order of magnitude, it was a correct solution to the problem and a marvellous piece of scientific insight.

Newton's Unanswered Questions

At this point I come to the end of my account of Newton's most solid discoveries, but no picture of his role as an explorer of the universe would be complete without some mention of his more tentative speculations. They reveal a mind reaching out in all directions to understand how the universe is put together

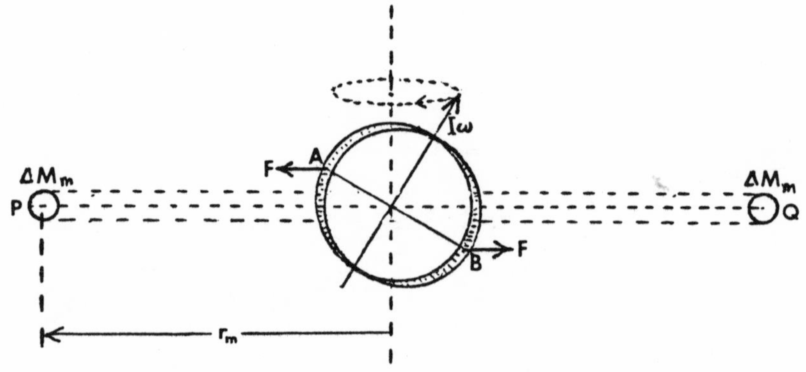

Figure 20. Mechanical basis of the precession of the equinoxes. [From French 697.]

and how it works. Newton's curiosity in these matters knew no limits, and his brilliant scientific imagination is always in evidence.

The *Principia* itself contains some fascinating examples. Several of them have a bearing on the ages of the Earth and the solar system. In his discussion of the heating of comets by the sun, Newton offers an estimate of how long it would have taken for the Earth to cool down if it had started out as a red-hot ball (*System* 522). His result was at least 50,000 years, far shorter than Baron William Kelvin calculated on the same basis two centuries later, but disturbingly long for any of Newton's contemporaries who might have been tempted to believe Bishop Ussher's date of 4004 BC for the Creation. A more definite conjecture concerned the persistence of planetary motions. Astronomical records made it certain that these motions had continued undiminished for thousands of years; clearly they experienced no detectable resistance. Was space simply a void? Newton was not sure, but certainly there could be no "gross matter." This fitted in well with what Newton had calculated about the Earth's atmosphere. He had discovered the theoretical law of an exponential decrease of air pressure and density with height (*Resisting Mediums* 297-98). Both would fall by a factor of a million for each 75 miles or so, and hence the space between the planets must be an essentially perfect vacuum (*Opticks* 367).

And yet ... . If there was no intervening medium, how was the force of gravity transmitted? In one of his most famous statements – a letter to Richard Bentley, D.D., on 25 February 1693 – Newton said:

> That one body may act upon another at a distance through a vacuum without the mediation of anything else ... is to me so great an absurdity that I believe no man who has in philosophical matters any competent faculty of thinking can ever fall into it. (*Correspondence* 3: 254)

Was the explanation to be found in an inanimate medium – the ether – or in the hand of God? Although Newton's mechanics seemed to be thoroughly deterministic, Newton himself had no doubt about the role of the Creator in getting it all started. How, otherwise, would we find the planets and all their satellites circulating in the same sense in almost the same plane? But he was much attracted by the idea of the ether, even though his pendulum experiments had quite failed to reveal it. I will return to this shortly. But, before I leave the *Principia*, let me mention one other wonderful speculation – concerning the distances of the stars.

Until close to Newton's time, the stars had been accepted as a fixed background to the motions of the Earth and the rest of the solar system. The idea developed that they might be bodies like our sun, but even through a telescope they still looked like luminous points, revealing nothing of their size. Newton found a way to tackle this problem (*System* 596). He noted that a prominent (first magnitude) star looked about as bright as Saturn. He knew how far away Saturn is; and also knew that we see Saturn by the sunlight that it scatters back towards us. Given that the intensity of light from a source falls off as the inverse square of the distance, he could calculate how far away a star like our sun would have to be to look as bright by direct radiation as Saturn does by reflected light. His result, expressed in modern terms, was about ten light-years, which is absolutely of the right order of magnitude!

At the end of the *Opticks*, in a series of what he called Queries, Newton set down a number of his final speculations about the physical world (339-406). In the first edition of the book, in 1704, there were sixteen of them, but in the third edition (1721) they had grown to thirty-one. Many, but not all, are concerned with optics: "Do not Bodies act upon Light at a distance, and by their action bend its Rays?" (Query 1). "Do not all fix'd Bodies, when heated beyond a certain degree, emit Light and shine; and is not this Emission perform'd by the vibrating motions of their parts?" (Query 8). "Are not the Rays of Light very small Bodies emitted from shining Substances?" (Query 29). "Do not several sorts of Rays make Vibrations of several bignesses, which according to their bignesses excite Sensations of several Colours, much after the manner that the Vibrations of the Air ... excite Sensations of several Sounds?" (Query 13). Most of these are accompanied by considerable amplification and discussion. It is clear that Newton wanted very much to incorporate a vibratory aspect into the description of light, but yet, not having quite recognized the existence of diffraction of light, he could not accept the notion that it was propagated as waves in a medium. Nevertheless, he came down on the side of there being an ether permeating all space and all bodies. We have seen something of this in his explanation of the "fits" of transmission or reflection in thin films. And in the Queries he offered further justification. In Query 18 he argued that the ether was needed to account for the transmission of heat across a vacuum; in Query 19 he postulated that the refraction of light required ether to be more dense in vacuum than in solids, for in his corpuscular theory the speed of light had to be assumed greater in (for example) glass than in air. Even in vacuum, however, the ether could be assumed so rare that it did not impede the planets. He deduced this (Queries 21/22) on the basis that the speed of vibrations in the ether had to be at least as great as the speed of light in vacuum and using the same formula as he had obtained for the speed of sound (Speed = Square root of the ratio Elasticity/Density). The next to last Query (number 30) seems to foreshadow Einstein: "Are not gross Bodies and Light convertible into one another, and may not Bodies receive much of their Activity from the Particles of Light which enter their Composition?" It sounds just like a qualitative statement of $E = mc^2$! Of course it was not, and could not have been; Newton's concern here was to compare the absorption and emission of light with various other chemical and biological transformations. But it is yet one more example of his boundless capacity for speculation.

The last Query (number 31) introduced a lengthy discussion in which Newton surveyed the whole range of the forces of nature. He began: "Have not the small Particles of Bodies certain Powers, Virtues, or Forces, by which they act at a distance, not only upon the Rays of Light for reflecting, refracting, and inflecting them, but also upon one another for producing a great Part of the Phenomena of Nature?" He then went on to discuss chemical reactions, surface tension phenomena, and the cohesion of solids, and speculated on the possible role of electric and magnetic forces in producing them. He expressed his belief in a fundamental uniformity of physical phenomena, saying: "And thus Nature will be very conformable to her self and very simple, performing all the great Motions of the Heavenly Bodies by the Attraction of Gravity which intercedes those Bodies, and almost all the small ones of their Particles by some other attractive and repelling Powers which intercede the Particles." It was a great vision, which he had done more than any person to make possible. But more than two centuries had to pass before science could develop far enough to make it a reality.

Concluding Remarks

In 1946, at Isaac Newton's own college of Trinity, Cambridge, the Royal Society of London held a celebration, delayed for four years by World War II, of the tercentenary of Newton's birth. The first main address at that gathering was given by Professor Edward Neville da Costa Andrade, a former research colleague of Ernest Rutherford and a devoted student and collector of Newton's works. He began his address with these words:

> From time to time in the history of mankind a man arrives who is of universal significance, whose work changes the current of human thought or of human experience, so that all that comes after him bears evidence of his spirit. Such a man was Shakespeare, such a man was Beethoven, such a man was Newton, and, of the three, his [Newton's] kingdom is the most widespread. (Royal Society 3)

To support this last claim, Andrade argued that the full richness of Shakespeare is partly lost in translation, that even music takes different forms in different cultures, but that the possibility of appreciating science is universal.

Although from one point of view Andrade's assessment is justified, it could not possibly be claimed that the numbers of those who appreciate Newton's science have ever approached, even remotely, the numbers who flock to see Shakespeare's plays or to hear Beethoven's symphonies. Nevertheless, in another sense, Newton's kingdom *is* the most widespread. Instead of the world of human experience, which for all its richness is limited to this planet, Newton's world was nothing less than the whole physical universe, and his exploration of it was, in my opinion, something the like of which the world has not seen before or since. I have never forgotten the thrill, as a fresh undergraduate at Cambridge, of knowing that I was walking where Newton himself had once trod. And, over the decades since that time, I have come to appreciate more and more the justification for the description of him (engraved on his statue in Trinity College chapel) as the man "who surpassed the human race by his genius." He was indeed incomparable.

NOTES

1 These stories of Newton's childhood come mostly from Stukeley; qtd. in Westfall 61-62.
2 From one of Newton's notebooks, about 1665-66. Reproduced in Newton, *Questiones* 468.
3 Voltaire was a great admirer of Newton, whose theories he promoted on the Continent against those of Descartes. The book is a remarkably good non-mathematical presentation of Newton's science – but Voltaire's brilliant mistress, the Marquise du Chastellet, to whom he dedicated it, went one better and translated the *Principia* into French!
4 Newton's *Optical Papers* contains the full text of both versions (*Lectiones* and *Optica*), with English translations.
5 See, for example, Berry, *Principles*, 37-38.

WORKS CITED

Berry, Arthur. *A Short History of Astronomy*. 1898. New York: Dover, 1961.
Berry, Michael. *Principles of Cosmology and Gravitation*. Cambridge and New York: Cambridge UP, 1976.
Cassini, Giovanni. *Philosophical Transactions of the Royal Society*, 16 (1686/87): 299-306.
Christianson, Gale E. *In the Presence of the Creator: Isaac Newton and his Times*. New York: Free Press, 1984.
Cohen, I. Bernard. *Isaac Newton's Papers and Letters on Natural Philosophy and Related Documents*. 2nd ed. Cambridge MA: Harvard UP, 1978.
______. "Newton." Vol. 10 of *Dictionary of Scientific Biography*. 16 vols. New York: Charles Scribner's, 1970-80.
Descartes, René. *Oeuvres de Descartes*. 12 vols. Paris: L. Cerf, 1897.
______. *Principia Philosophiae*. 3 vols. Descartes, Vol. 8 of *Oeuvres*.
French, A.P. *Newtonian Mechanics*. New York: W.W. Norton Co., 1971.
Gjertsen, Derek. *The Newton Handbook*. London and Boston: Routledge and Kegan Paul, 1986.

Hauksbee, Francis. *Philosophical Transactions of the Royal Society*, 27 (1710/12): 196-98.
Herivel, John. *The Background to Newton's* Principia. Oxford: Clarendon Press, 1965.
Huygens, Christiaan. *Philosophical Transactions of the Royal Society*, 4 (1669): 335-38.
Newton, Isaac. Vol. I of *The Correspondence of Isaac Newton*. Ed. H.W. Turnbull. 7 vols. Cambridge UP, 1959-77.
———. "Letter to Henry Oldenburg." 19 March 1671/72. Newton, *Correspondence* I: 121-22.
———. "Letter to Henry Oldenburg." 13 April 1672. Newton, *Correspondence* I: 137.
———. "Letter to an Unidentified Friend." 23 February 1668/69. Newton, *Correspondence* I: 121-22.
———. *Motion of Bodies*. Book I of *Principia*.
———. *Motion of Bodies in Resisting Mediums*. Book II of *Principia*.
———. *The Optical Papers of Isaac Newton*. Ed. Alan E. Shapiro. 3 (projected) vols. Cambridge and New York: Cambridge UP, 1984, Vol. I.
———. *Opticks*. 1730. New York: Dover, 1952.
———. [*Principia*] *Sir Isaac Newton's Mathematical Principles of Natural Philosophy and his System of the World*. Ed. Florian Cajori. Trans. Andrew Motte. 2 vols. Berkeley: U of California P, 1934.
———. *Questiones Qu'dam philosophiae. Certain Philosophical Questions: Newton's Trinity Notebook*. Ed. J.E. McGuire and Martin Tamny. Cambridge and New York: Cambridge UP, 1983.
———. *System of the World*. Book III of *Principia*.
The Royal Society (Great Britain). *Newton tercentenary Celebrations*. Cambridge: Cambridge UP, 1947.
Stukeley, William. *Memoirs of Sir Isaac Newton's Life*. London, 1752.
Voltaire. *The Elements of Sir Isaac Newton's Philosophy*. 1738. New York: Frank Cass, 1967.
Wallis, John. *Philosophical Transactions of the Royal Society*, 3 (1668): 307-10.
Westfall, Richard S. *Never at Rest: A Biography of Isaac Newton*. Cambridge: Cambridge UP, 1980.
Wren, Christopher. *Philosophical Transactions of the Royal Society*, 3 (1668): 310-3312.

From White Dwarfs to Black Holes: The History of a Revolutionary Idea

WERNER ISRAEL

Although it is hard today to remember a time when black holes were not a part of everyday discourse, only twenty years ago no one had ever heard the term "black hole" in an astronomical context. It was coined by John Wheeler of Princeton in an after-dinner talk to the American Association for the Advancement of Science in New York on 29 December 1967. The name and its connotations immediately captured the public imagination. Books on black holes are now to be had in every bookstore, though to find them it may be necessary to search the shelves on mysticism and the occult.

The notion of a black hole, in contrast to the name, is a very old one. It goes back two hundred years, as far as we have any printed record, to the time of Pierre Laplace. And it may well be as old as the theory of gravitation itself. As a natural offspring of his two most famous hypotheses, universal gravitation and the corpuscular theory of light, it is just the sort of idea that might have occurred to Isaac Newton himself during an idle moment in the bath. If light consists of particles, and if all particles are subject to gravity, then gravity should have the power to bend and decelerate rays of light. From there, it is but a short step to contemplate the existence of stars whose gravity is so strong that their light is held back and they appear to be invisible.

Speculations of this kind may be found in the late eighteenth-century writings of John Michell in England, Pierre Laplace in France and Johann Soldner in Germany. They were quickly consigned to limbo when the rise of the wave theory of light – especially after the discovery of interference by Thomas Young in 1801 – removed any obvious reason to believe that light should be influenced by gravity. A hundred years later they were briefly resurrected when the 1919 eclipse expedition confirmed Einstein's prediction of the bending of light in the sun's gravitational field.

From the earliest days up to the 1930s, speculation about black holes almost always envisaged monster objects, about as large as the Earth's orbit, that weigh one hundred million times as much as the sun and have the density of water. That black holes might exist which were not much heavier than the sun was an idea that almost no one took seriously before the 1960s. Such objects would be only a couple of miles across, and the corresponding densities seemed (in the words of Sir Oliver Lodge) "beyond the range of rational attention." This dismissive attitude now appears strange, since by the 1930s there were already cogent theoretical reasons to believe that such black holes should exist as tombs of massive burnt-out stars. But this idea was to be fiercely resisted for more than thirty years.

Although physical scientists have generally shown surprising resilience in adapting to the most outlandish concepts, from the relativity of time to the curvature of space and the wave-like aspects of matter, two conjectures in our century were long stultified by universal scepticism: continental drift and the possibility of solar-mass black holes. Perhaps it is not coincidental that these disparate ideas have a disturbing element in common: both call into question our deep-rooted instinctive belief in the permanence and stability of solid matter.

The curious story of black holes is the theme of this article. But before delving into history, it will be useful to gain some perspective with a brief review of black holes and their place in the universe according to our present understanding.

As everyone knows, a black hole is a region of space in which gravity is so strong that matter and light cannot escape to the outside. The surface of a black hole is not solid; an astronaut passing through the surface of a black hole will encounter nothing but emptiness. In fact, if the black hole is sufficiently large – for instance, one of the monster black holes now suspected to lurk in the cores of violently active galaxies – nothing at all will seem unusual. But in less than ten minutes, approaching the centre, one will begin to feel head and feet being stretched apart and chest squeezed by growing tidal forces. Attempts to flash a distress signal to the outside are doomed to fail, because the light is dragged *inwards* by gravity, though not as fast as the flashlight that emitted it! Below, at the centre, looms a Thing that is beyond the power of present-day science to describe. We can say only that here our ordinary notions of space and time break down utterly and, as a cloak for our ignorance, we call it a "space-time singularity." The astronaut is unable to see this Thing, because it lies in the future: it cannot be experienced before it is reached. Well before then the body will have been torn apart and crushed by overwhelming tidal forces.

Let us turn from this pitiful spectacle to consider the place of black holes in the astronomical universe. Of the stars visible to the naked eye, none is a very densely packed object. The very fact that they are luminous implies that they are kept distended and diffuse by the outward pressure of their heated gases. The average density in the sun is about one and one-half times that of water; in red giant stars it is much less than air. However, we are now aware of three types of object in which a large amount of mass is packed into a relatively small volume. In *white dwarf stars*, a mass equal to the sun's is packed into a volume comparable to the earth's. A matchbox of this material would weigh a ton. Even this is trifling compared with conditions in *neutron stars*, where the bare nuclei of atoms are jammed together, and the density approaches that inside a nucleus: about a million billion times that of water. At the centres of *black holes* matter may actually have been squashed out of existence, with only its gravitational field (like the grin of the Cheshire cat) remaining.

These three types of object are all believed to be corpses of burnt-out stars. To understand how they arise, we should look briefly at the life-cycle of a star. In our galaxy, each year about twenty new stars are born out of the interstellar gas. The primordial star is a diffuse cloud that slowly contracts under its own gravity. Compression of the gas heats the cloud. When the temperature near the centre has risen to about a million degrees, nuclear reactions are ignited. The contraction then stops, and the cloud begins life as a star, steadily burning its nuclear fuel. The sun has enough fuel to last more than ten billion years. Heavier stars need more heat to keep themselves distended against gravity, and burn up their fuel supply considerably faster.

The change of chemical composition as the star first uses up its hydrogen and then begins to burn helium, produces gradual changes in the star's physical structure. Computer calculations show that the star will swell and grow brighter while its surface becomes cooler (i.e., it will grow redder). The star will evolve into a red giant. The time-scale of this evolution is several billion years for the sun, and about a thousand times less for a star a thousand times as luminous (which would make it about ten times as massive).

After the red giant stage the evolution becomes too rapid to follow with present-day computer technology and one must rely on plausible guesswork. When all combustible fuel has been exhausted, the star has no option but to contract slowly under its own gravity. Heat derived from compression of the gaseous material will enable the star to keep shining for another few million years. The core will become so hot that the intense radiation pressure will drive off the outer layers, forming a spherical shell of hot gas some light-years from the star. (The Ring nebula in Lyra is a picturesque example of such a shell.) In the case of the sun, or any star of comparable or smaller mass, the subsequent evolution is uneventful, with the core slowly contracting to about the size of the Earth, ending as a white dwarf slowly cooling to invisibility.

No such peaceful end is possible for a star several times heavier than the sun. As long as a store of heat was available, the star could hold up its weight by the thermal pressure of its hot gases. But this thermal contribution declines as the star cools. A moment arrives when the almost cold material cracks and disintegrates under its own weight. Even the hardest terrestial granite caves in under a weight of more than ten million atmospheres. White dwarf material is much stiffer, but even it can support only 10^{20} atmospheres before it, too, succumbs. The star collapses catastrophically. The sudden compression heats the core to nearly a billion degrees, releasing previously untapped nuclear resources and blowing off the outer layers. Astronomers on earth, if they are favourably positioned, will observe a supernova explosion like the highly publicized event seen earlier this year in the Large Magellanic Cloud.

In 1064 Chinese astronomers recorded the most famous explosion of this kind. Its debris is now observable as the Crab nebula. There is still a great deal of activity in this remnant; in particular, it is an extended source of X-rays, one of the strongest in the sky. A central hot spot in the X-ray emission has been identified with a star that blinks on and off thirty times per second – a pulsar – which is undoubtedly the core of the star that exploded. Pulsars are believed to be rapidly spinning neutron stars. The pulses are probably a lighthouse effect, seen whenever a narrow beam of radiation from the magnetic pole (non-aligned with the spin axis) crosses the line of sight.

Neutron star material is much harder even than white dwarf material (it can bear up to 10^{30} atmospheres), but it is not infinitely hard. If the collapsing core that remains after the outer layers have been blown off is still as heavy as three or four solar masses, the collapse will crash through the neutron star stage. No barrier is left to prevent it from plunging all the way to zero volume, forming a space-time singularity surrounded by a black hole.

At present we have at least two good observational candidates for star systems containing a black hole. Both involve strong X-ray emission associated with a blue giant star which is orbiting an invisible, compact and very massive object. The invisible star – the presumptive black hole – appears to be "breathing in" the atmosphere of its giant companion. It is thought that gas collects around the black hole in the form of a disk and gradually spirals in, while viscous effects raise it to temperatures of one hundred million degrees, producing the observed X-ray emission.

Let us now enlarge our field of view from individual stars to entire galaxies. We find a web of circumstantial evidence pointing to the presence of giant black holes in galactic nuclei, with masses ranging from a million to a hundred million solar masses. Observation of the central regions of our own galaxy at radio, infrared, X-ray, and even gamma-ray wavelengths, reveals a hot spot precisely at the galactic centre. There have been suggestions – but no firm evidence to date – that this may be a manifestation of accretion by a million-solar-mass black hole at the galactic centre.

Much stronger activity has been observed in the nuclei of many other galaxies. Radio images show spectacular jets streaming out of their centres, often associated with intense X-ray and infrared emission. The mechanism of jet formation is not well understood, but is thought to involve some complex interaction between a central black hole and a surrounding disk of ionized gas.

Nearly all of these spectacular phenomena would still be hidden from us without the techniques of radio, X-ray, and infrared astronomy that were developed after World War II. The pre-war astronomer, equipped solely with telescope and spectroscope, saw the universe only through the narrow optical window, and it seemed a much quieter place. Pre-war astronomers were, not surprisingly, a conservative breed who saw their task as the construction of an orderly framework for the peaceful landscape which their observations revealed. It was this generation that had to cope with the trauma of a chasm gradually opening under their feet – first, with the dawning realization around 1914 that stars might exist tens of thousands time denser than water; then, with theoretical evidence in the 1930s that even this might not be the ultimate limit to which matter could be compressed.

As we struggle to keep abreast of the explosion of current information it is fair to ask why a harried contemporary scholar should be concerned with the naïve thinking and (in retrospect) dumb mistakes of our predecessors in 1914. There are several ways to answer that. The pragmatic answer is that history is the tower of experience. Knowing a certain amount of history may help us to avoid repeating the old mistakes. It is true that in science important mistakes are never made twice. The mistakes that count are the ones at the frontiers and the frontiers are constantly moving outward. But it is easy enough to fall into the same *kind* of mistake, and in this, it seems, we never fail, generation after generation.

What was the mistake of young astronomers in 1914? They scrutinized the most modern and advanced astronomical theory of the day. Because it was so beautifully ingenious, plausible, and tied together so many loose ends, they accepted it as received truth, thereafter not to be questioned. It was this almost subliminal acceptance, more than anything, which set back recognition of the true nature of white dwarfs by more than a decade.

The hot new theory of 1914 was a scenario for stellar evolution proposed by two of the astrophysical pioneers of our century, Ejnar Hertzsprung working in Leiden and Henry Norris Russell of Princeton. Ironically, both Hertzsprung and Russell were sober and careful men who advanced their hypotheses with reserve, and increasingly voiced misgivings as counter-evidence began to accumulate in the 1920s. But by then the giant and dwarf evolutionary theory (as it was called) was firmly entrenched as the conventional wisdom.

Hertzsprung and Russell had plotted all stars for which they knew the distance – in 1914 this essentially meant the nearest stars – on a colour-luminosity, or H-R, diagram. Figure 1 is a modern version of this diagram for the nearest stars. Intrinsic brightness or luminosity is plotted vertically on a logarithmic scale. The horizontal scale shows surface temperature or colour. Red stars, with surface temperatures of 3,000°C are on the right, yellow stars like the sun (surface temperature 6,000°) in the middle, white stars like Sirius (surface temperature 10,000°) toward the left, and blue stars on the extreme left.

Examining their plot, Hertzsprung and Russell made a fundamental observation. They noticed that almost all the stars in our vicinity fall into two distinct classes. The majority are on a narrow band (nowadays called the "main sequence") while a smaller group occupies a broader band higher up. Two stars having the same colour and hence surface temperature should radiate the same amount of energy per square foot. Therefore, if one is 10,000 times dimmer than the other, its surface must be 10,000 times smaller. This warranted referring to stars in the upper and lower bands as giants and dwarfs respectively.

The theory of Hertzsprung and Russell was an ingenious attempt to explain these two bands as two stages in the evolution of each star. Before going into this, I shall briefly explain the present-day interpretation. All stars begin their lives with the standard chemical composition of the interstellar gas from which they originate - about three parts hydrogen, one part helium, and a smaller admixture of heavy elements - determined by nucleosynthesis in the first few minutes of the big bang. According to the theory of stellar structure developed in the 1920s, this means that all stars begin on the same starting line in the H-R diagram - the main sequence - with the heavier stars being the more luminous, i.e., falling higher up on the line. As their composition changes through nuclear burning, all will evolve off the main sequence. This evolution will proceed much faster for the heavier, brighter stars than for modest fuel-consumers like the sun. Now, most of the stars in our neighbourhood are no heavier than the sun, and we find them still close to the main sequence, from which they have not yet had time to evolve appreciably. A relatively small number of objects that are appreciably heavier have managed to evolve to the red giant stage and now occupy the upper band. This is the present view of the evolutionary significance of Figure 1.

A very different picture was painted by the giant and dwarf evolutionary theory (which, it must be remembered, was proposed well before anything was known about nuclear reactions). Each star was envisaged as beginning as a distended, cool globe of gas, a red giant. Compression and heating of the gas as the star slowly contracts under its own gravity should raise the surface temperature and the star will evolve toward the blue end of the main sequence. At this point the density is comparable with water and (so the argument ran) the material should be in the condition of an incompressible liquid. Since fur-

Figure 1. Hertzsprung-Russell (H-R) diagram showing absolute magnitudes of stars as a function of their spectral class, related to temperature. [From The Cambridge Atlas of Astronomy, *original rights,* Le Grand Atlas de l'Astronomie, *Jean Audouze and Guy Israel, Paris: Encyclopaedia Universalis France.]*

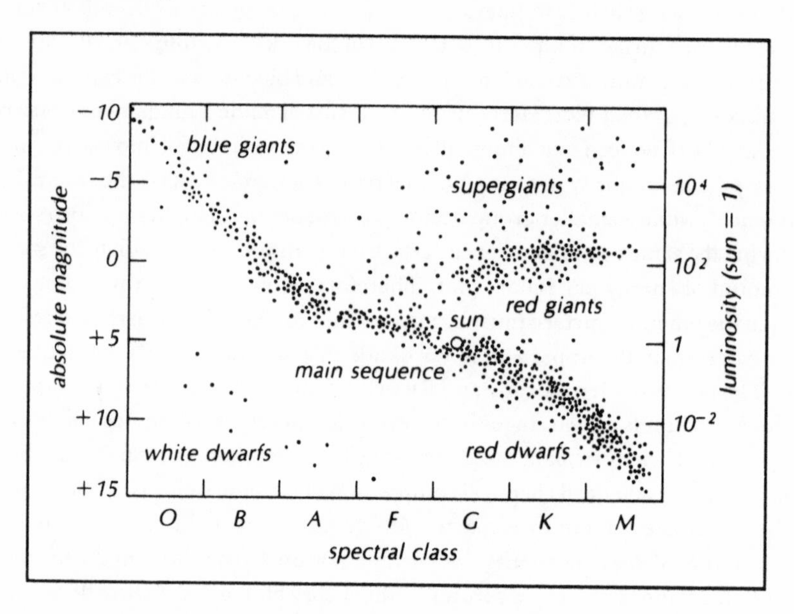

ther contraction appears impossible, the star has no option but to cool off gradually. It was supposed that in this process of cooling the star would evolve down the main sequence to end as a red dwarf.

The idea that a star could evolve along the main sequence now appears almost absurd, since we know that different positions correspond to different masses. But it caused no qualms in 1914, when data on stellar masses were sparse and a mass-luminosity correlation was not yet suspected.

Nevertheless, there was one small piece of evidence which should have – and did – give some cause for concern. Figure 2 shows the original H-R diagram that Russell prepared in 1914. There is a good deal of scatter – data on stellar distances were not yet too reliable – but one can clearly distinguish the main sequence and the giant branch. One point stands out jarringly in the lower left-hand corner: a white star of abnormally low luminosity, known as 40 Eridani B. A star in this position seemed an absurdity: 10,000 times fainter than Sirius, and of the same colour, its surface area would need to be 10,000 times smaller, and its radius one hundred times less, which would have made it only about twice as large as the Earth. Yet its mass was known (from its perturbing effect on the orbit of a companion star) to be at least a quarter of the sun's. The inferred density appeared too ridiculous for serious discussion.

Figure 2. Original H-R *diagram prepared by H.N. Russell in 1914.*

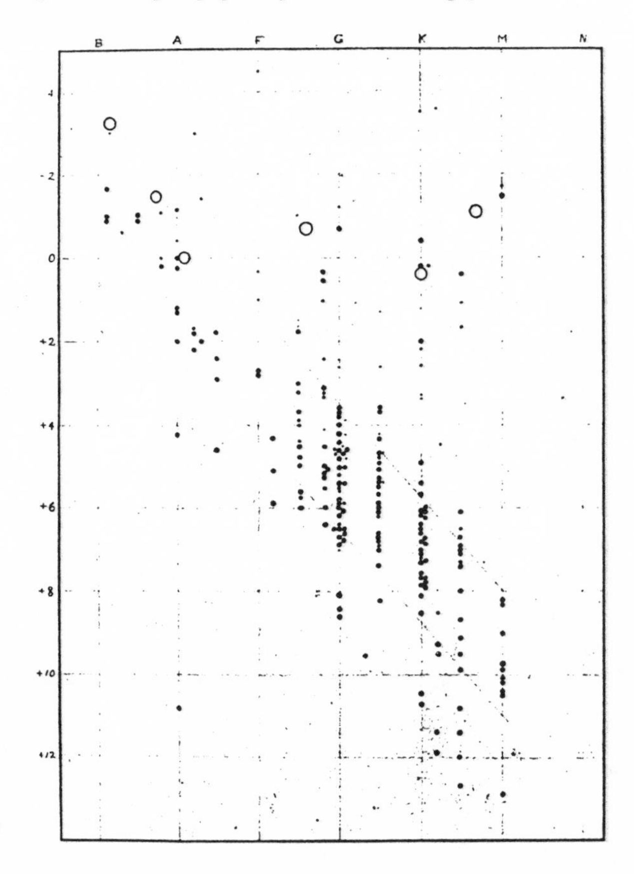

Russell in 1914 was inclined to the view that the anomaly might be attributed to human causes, that the vagrant point really belonged over on the right with the red dwarfs. Perhaps, just once, the usually dependable people at the Harvard College Observatory had nodded while scanning the spectrum and classifying the colour of this star.

This explanation began to look distinctly strained when 40 Eridani was joined by a second freak. In December 1915, Walter Adams at Mount Wilson Observatory announced that he had finally secured a spectogram of the faint companion of Sirius and found it to be a white star. Glare from the primary makes this observation about as easy as making out the colour of a glow-worm on the headlamp of an oncoming roadster, and a successful outcome had to await recession of the companion to its furthest separation from Sirius, which occurs only once in fifty years. By 1919, three of these freaks were known which nobody knew what to make of, although there was unanimous agreement that densities of 50,000 gm per cm^3 were "impossible" and "absurd."

The situation was ironic, since the essential clue to the paradox had been furnished by the theoretician James Jeans in 1916, and sat unnoticed under everyone's nose for eight years! At a meeting of the Royal Astronomical Society in London, the Cambridge astronomer Arthur Eddington had presented the first of his pioneering papers on physical conditions inside a gaseous star. In the ensuing discussion, Jeans pointed out that the value 54 which Eddington had adopted for the mean molecular weight of the material (close to the atomic weight of iron) was certainly too high, since the atoms would be disrupted by violent collisions at the temperatures (exceeding ten million degrees) prevailing inside the star. Even a star of pure vaporized iron would have a mean molecular weight of no more than 2, since the total weight of an atom would be shared between twenty-seven freely moving particles, twenty-six electrons and one nucleus.

Eddington accepted this criticism in his subsequent work. But it somehow escaped notice that disruption of the atoms would also enormously reduce the effective *size* of the particles. This means that matter in stars can go on behaving as a compressible gas at densities far beyond the point where the same matter on Earth would have become liquid. In general, matter ceases to be gaseous when its particles become jammed and can no longer move freely. On Earth, this happens at densities comparable with water, when atoms become jammed; in stars, it will not occur before bare nuclei are jammed, i.e., at nuclear densities. Thus there is nothing inherently absurd in the notion that stellar material should be compressible to densities of 100,000 or even higher.

Why was this simple point, so obvious in retrospect, a blind spot to the astronomers for eight years? As Eddington later admitted, it was at least in part their attachment to the giant and dwarf theory and the associated preconception that main sequence stars could not be gaseous.

When the penny finally dropped, it was by a very indirect route. In 1924, Eddington worked out a theoretical relation between the masses and luminosities of stars with a given chemical composition. Since he had to use Boyle's law for a perfect gas in his derivation, he expected only the tenuous giant stars to fall on this theoretical curve. One can imagine his amazement to find the sun, and even stars with the density of iron, on the curve as well. After this, the pieces fell quickly into place. The meeting of the Royal Astronomical Society at which he presented these new results and insights was a historic turning point in astrophysics.

It is a general rule in science that the solution of one problem leads straight to another. In 1926, Eddington encountered a riddle which has come to be known as Eddington's paradox. He posed the question, What will happen when a white dwarf finally cools to absolute zero? It was well established that heat is a manifestation of the random motion of particles. It was therefore expected that at absolute zero this random motion should cease, and the particles would stop colliding. But, since it was these very collisions that had kept the atoms broken up, there would now be nothing to prevent nuclei from recombining with electrons as complete atoms. The star would have to expand at least 10,000-fold to accommodate the much larger volume of its atoms. In the process it would have to do an enormous amount of work in pushing its material outward against gravity. Where would it find the energy to do this? Its available fuel has been exhausted and it has no other resources. "Imagine," said Eddington, "a body continually losing heat, but with insufficient energy to grow cold!"

In 1926 the new quantum mechanics had just been formulated, and the Cambridge theoretician Ralph Fowler pointed out that it offers a way out of Eddington's paradox. According to quantum mechanics, heat is not the only form of random motion; there is a second, purely quantum, form that lingers in the electrons of a body even at absolute zero and can never be lost. It can be understood as a consequence of Heisenberg's uncertainty principle, which states that the product of the uncertainties in the momentum and position of any particle can never be smaller than 10^{-34} MKS (metre-kilogram-second) units. Thus, the momentum of an electron confined to a given volume can never be exactly zero, since then it would have no uncertainty! Some random motion must be retained, even in a cold body.

On Earth, this random motion of electrons provides an irreducible pressure that gives terrestial atoms and solids their appearance of rigidity. In white dwarfs, it provides the collisions that keep the atoms disrupted and also the pressure that holds the star up. At one stroke, all the loose ends had been satisfyingly sewn up. Or so it appeared. In any event, this is where matters lay for three years.

The first hints that Fowler's explanation was not yet the last word came, coincidentally, at the time of the Wall Street disaster. Imagine an armchair experiment in which we have a white dwarf of moderate mass, say half a solar mass, and we add a small amount of extra mass. The star will contract under the extra load. As the volume to which they are confined shrinks, the electrons have their positional uncertainties reduced, hence their momentum uncertainties must increase to compensate. Their pressure will rise until it balances the extra weight and the contraction stops. Thus, the heavier star will settle into a new equilibrium at a smaller radius.

But this recovery is no longer assured if we continue to load mass onto the star indefinitely. At some stage the electrons will approach and stick at the speed of light. Although their momenta are now very large, the electrons can no longer transport them fast enough to provide the pressure support needed to hold up the star. The star faces an implosion crisis.

The first clear recognition of this crisis came in June 1930 from the Indian astrophysicist Subramanyan Chandrasekhar, not yet twenty at the time, and en route in mid-ocean to Cambridge for graduate study. His calculation showed that no white dwarf heavier than 1.44 solar masses would be able to support its own weight.

The following year the young Soviet physicist Lev Landau independently did the same calculation. So appalled was he at the outcome that he wrote, "For masses greater than 1½ solar masses there exists in the whole quantum theory no cause preventing the system from collapsing to a point. ... As in reality such masses exist quietly as stars and show no such ridiculous tendencies, we must conclude that all stars heavier than 1½ solar masses certainly possess regions in which the laws of quantum mechanics are violated." At this point, Landau was prepared to cast all of physics back into the melting-pot to evade his own "ridiculous" conclusion.

Eddington's attitude to the new developments was equally sceptical. Addressing the Royal Astronomical Society in January 1935, he said, "I don't know if I shall escape from this meeting alive, ... but I think there should be a law of nature to prevent a star from behaving in this absurd way." He proceeded to argue that the combination of quantum mechanics and special relativity used in deriving Chandrasekhar's mass limit was "an unholy alliance" and unsound.

Chandrasekhar's analysis was in fact straightforward and entirely correct. But so deep-rooted was the general resistance to the idea of gravitational collapse that it was universally shunned and soon forgotten, even though Robert Oppenheimer and his students at the University of California had published seminal papers supporting it just before the war. Its fortunes did not revive until 1963, when the discovery of quasars suddenly made it attractive and even necessary to look to black holes as a possible source of their prodigious energies.

One cannot help speculating on psychological reasons for the protracted resistance to the idea. The key lies, I think, in the emotive words "ridiculous" and "absurd." Even after the density of white dwarfs was conceded, there remained a deep repugnance to the idea that matter could be crushed into nothingness. If atoms could be crushed and broken up, then surely at some more basic level, perhaps that of the nucleus, one would encounter the infinitely hard objects that Lucretius had spoken of two thousand years before. Lucretius himself had warned of the consequences of allowing any compromise on this issue: "The ground will fall away from our feet, its particles dissolved amid the mingled wreckage of heaven and earth. The whole world will vanish into the abyss ... . At whatever point you allow matter to fall short, this will be the gateway to perdition."

The last two decades have seen a complete about-face in our thinking about the compressibility of matter. The gradual acceptance of gravitational collapse, the discovery of the quark structure of nuclei and other experimental and theoretical advances have left little reason to believe that the ultimate particles of matter are anything more than mathematical points. Cosmologists today talk routinely about 10^{78} gm/cm^3, the density of the universe 10^{-35} seconds after the big bang. Perhaps the ultimate limit of compressibility is reached only at 10^{94} gm/cm^3, when quantum fluctuations disrupt the geometrical fabric of space itself, and the very notions of volume and density become meaningless. But here we trespass beyond the borders of present knowledge and understanding.

The Newtonian Contribution to Our Understanding of the Computer

STEPHEN SMALE

Isaac Newton made a continuous mathematical model of a discrete universe using differential equations to explain how things in that universe move. Mathematicians can look at the computer today in the same way that Newton looked at the world in the seventeenth century to gain some understanding of "the machine." One can try to understand the computer in a way similar to the way a physicist understands the world, a different way from the engineers' approach. In order to understand the laws of computation one can design a mathematical model using continuous mathematics and calculus to lead to a deeper understanding of the process of computation.

The model involves going from the discrete to the continuous. An example which helps explain the difference between the concepts of discrete and continuous is that of clock faces. An old watch with a traditional face is a continuous object. The hands move continuously around the face and the whole spectrum of "possible" time. One gets a continuous picture rather than a discrete picture of the time. A modern watch using digits shows only the present minute. As time passes it will show you another and another, but each minute is separate and discrete.

The world before Isaac Newton was perceived as fundamentally discrete in nature. Thomas Kuhn's book *The Copernican Revolution* discusses very well the background of science when Newton began to work. The understanding of the universe was based on the theory of atomism which was first promulgated by the Greeks, Democritus and Lucretius. Atomists described the matter in the universe as being composed of a finite – discrete – number of particles. Each particle is indivisible so they are called the elementary particles of matter. It was this picture of the nature of matter and the universe which was generally accepted when Newton began his work. Newton believed that the universe was made up of a finite number of building blocks. He used the word corpuscular and he believed in a corpuscular universe.

On the other hand, Newton's vision of the universe was based on the geometry of the Greeks with its notions of curves and lines, which are continuous objects. There is nothing finite at all about a smooth, continuous curve.

The resolution of the problem of the discrete universe and his continuous mathematical models was a crucial step for Newton. According to Thomas Kuhn this seeming contradiction delayed the publication of Newton's *Principia*.

Real numbers are crucial to Newton's vision of a continuous universe. They are the numbers representable by infinite decimal expansion. Real numbers give the picture of a continuous set of numbers between zero and one. This concept is basic to an understanding of how Newton reinterpreted the world. The success of real numbers in his continuous picture of the universe contrasted greatly with his picture of the world as being finite. Newton saw the Earth as made up of a lot of particles; each has gravity, is affected by the others, and has different forces pulling upon it in a very complex situation. He resolved this dilemma by calculating the forces exerted on the accumulation of particles of the Earth in this finite picture. He doubled the number of particles he considered in his calculation, filling out the picture more and more densely and making one calculation after another. He did it again and again, each time making a new calculation. Since *mathematically* the number of particles goes to infinity, one can think of the Earth analogously as a single particle located at the centre of an infinite universe. Kuhn, in the *Copernican Revolution*, says: "In 1685 [Newton] proved that, whatever the distance to the external corpuscle, all the earth corpuscles could be treated as though they were located at the earth's centre. That surprising discovery, which at last rooted gravity in the individual corpuscles, was the prelude and perhaps the prerequisite to the publication of *Principia*" (258). And Kuhn adds: "At last it could be shown that both Kepler's Law and the motion of a projectile could be explained as the result of an innate attraction between the fundamental corpuscles of which the world machine was constructed" (258).

Once Newton had devised that mathematical framework he could use the tools of calculus to understand the motions of the planets. Calculus is the mathematics of ordinary differential equations, on the space of states of physics. The "states" in physics describes the position of a particle with its velocity, or the positions and velocities of a number of particles. The basic laws of physics define differential equations on the space of states which govern how a state moves in time. For example, one might want to calculate the path of a planet with a specific velocity. Newton's Law, which was that force equals mass times the acceleration ($f = ma$), explains it in mechanical terms. But a differential equation explains it in mathematical terms, such as dx/dt equals some function of x, so x is a state and x will move in time. This equation says that motion is unique when one knows where the particular state is at time zero. In some sense this idea is the principal one that inspired the *Principia* and the way Newton calculated, for example, the ellipses of the planets. He explained the ellipses of

Kepler in terms of differential equations using the laws of gravity.

Kuhn says: "The construction of Newton's corpuscular world machine completes the conceptual revolution that Copernicus had initiated a century and a half earlier" (261). But Kuhn emphasizes the fact that Newton was assuming a corpuscular, atomistic, discrete world. And there were things that Newton did not deal with sufficiently in the *Principia*, things which have to do with partial differential equations. These are the laws by which fluids and solids move, the laws of motion of terrestrial objects in general. Newton attempted to deal with these matters in his book, but the attempt was successful only in outline, and it was left to such mathematicians as Daniel Bernoulli, Leonhard Euler and J.L. Lagrange to deal with these matters directly. They carried out the programme that was implicit in *Principia* by formally discovering the partial differential equations that show how fluids move. Euler's equations of fluid motion are an example.

Alan Turing was a very interesting and tragic figure, who about 1936 formalized our notion of a computer in a mathematical model which provides the theoretical foundation of digital computer science. His great contribution to computer theory is known as the Turing machine.

A Turing machine is not a stationary object with moving parts and levers and buttons. It is a "machine" on paper, a theoretical model. (An example of a machine which is a paper idealization is the flow chart, which depicts the sequence of work in a machine, the programme of a machine used for solving problems and scientific computations.) A simplified, idealized way to understand how a digital computer works might be to picture it as an input tape with information encoded on it consisting of either a zero or a one. As it moves through the computer a counter moves back and forth according to a system of instructions. What is on the tape when the process is completed is called the output of the machine. One sees in this picture of the Turing machine the property of discreteness. There is a long string of zeros and ones, and the machine can encode a lot of information because of the many possible combinations, but it is a very discrete process in the sense that the input of the machine consists of the limited set of numbers that I have been talking about. Aho, Hopcroft, and Ullman discuss the random-access machine in a way which is much closer to the way we think of the computer today. But even in random-access computers one thinks of the inputs as discrete numbers and they carry out the same input/output functions as Turing described.

I am quite critical of this idea of the computer as a finite instrument. Turing modelled his theory on logic which is a very specific and narrow part of mathematics. This has kept it away from the mainstream of mathematics and hindered its development. John von Neumann, the mathematician who built some of the first computing machines, writes frequently about the foundations of computer science and the general and logical theory of automata. In his opinion, "We are very far from possessing a theory of automata which deserves that name." He said the reason for this lack was that logic has very little contact with the continuous concept of a real or complex number, that is, with mathematical analysis. He went on to say: "A detailed, highly mathematical and more specifically analytical, theory of automata and of information is needed."

Computer scientists still rest their theories on the discrete idea expressed by Turing machines. Why I object to that, and why von Neumann expected more, has to do with scientific computing, which is the most complex use of the computing machine. Like differential equations, scientific computation addresses especially complex problems, such as predicting weather and providing aerodynamic models. Numerical analysis, the theoretical side of scientific computation, provides some basis for solving differential equations using algorithms. Some scientific people – engineers in scientific computation or theoreticians of numerical analysis – have very little respect for the Turing machine. The Turing theory is seen as old-fashioned and limited and remote from the kind of explanation they need. There is also a move away from the use of calculus in the discrete mathematics of computer science. What this has caused is an unhappy lack of unity between two disciplines, computer science and numerical analysis, two subjects which should be intimately related. What I would like to see is a programme for changing the foundations of computer science theory in the same way that Newton changed physics in the *Principia*, to idealize models for the computer which will relate to the algorithms of numerical analysis, and to explain mathematically the use of the idealization for scientific computation.

What I want to suggest is a different model for computer theory based on the idea of using real numbers. It will be an idealization just as the Turing machine model is an idealization, and it provides a different model to explain the same phenomena. In physics one has the Newtonian picture, a mechanical picture of the universe, but one has also a relativistic picture, and a quantum picture. Relativity is useless to explain most of engineering, which is based on Newton's mechanics. Different models explain different aspects of the same physical reality. One should not look for a single model of the computer, but rather look for models which will help us understand scientific computation on the one hand and classical computer science on the other.

Why is there this dichotomy between the two? What makes the computer discrete? Digital machines make calculations using data represented by the digits between zero and one. Machines have a great deal of precision but it is still a finite precision: they will enter numbers with accuracy, 10^{-8}, for example. Translated into this picture it means that they will put in, between zero and one, 10^8 numbers, about a hundred million. The computer will allow us to enter without doing any special programming a hundred million numbers in this interval. What this means is that while there is a discrete, finite, picture in terms of the input, the numbers that are inputted into the computer approximate the real numbers very well. To extrapolate the input from a finite number to a continuous picture of the inputs, to a real number input, is not difficult. It is an idealization, but it is a plausible idealization. We know that the universe is a discrete universe, and the laws of differential equations – Newton's mechanics – are still adequate to explain how matter moves within this discrete universe. In the same way I would suggest that we can understand at least one major facet of the computer better if we smooth over the discrete set of numbers to make a continuous set, as Newton did when he expanded a discrete particle theory to describe a continuous universe.

One can describe a machine in very abstract mathematical language but I will do it first in a simple way. I will start by describing a "tame" machine by means of the real numbers. Those of you who have studied computer science will say "that's nothing new," but let us see how a general foundation for computing functions over the real numbers develops.

In the domain of discrete mathematics, computer scientists have systematized the notion of a machine into recipes for solving discrete problems. What one hopes is that that could develop into a way of systematizing recipes for mainstream mathematics. An example of this is the fundamental theorem of algebra. You can imagine a function being given, say, z into some polynomial $p(z)$ by such a machine (theory) where we do not allow any division. We can imagine z going into $p(z)$ by addition, subtraction, and multiplication. That describes the polynomial. Is there some point which goes into zero under this particular machine? There is and it is given by the fundamental theorem of algebra. One has to allow things that are called complex numbers. How can one find the solution to the equation $p(z) = 0$, where p is the polynomial $az^2 + bz + c = 0$? This would be a polynomial of degree 2. Then one can study polynomials of degree d which look like az^d plus lower terms. What is the nature of the ways of finding the solution to this equation? A theorem which can be proved is that these machines must have the number of branches (or "if thens") proportional to the degree of the polynomial. In other words, to solve polynomial equations which are of higher degree requires more and more complexity. One needs more and more branches to solve the problem, which is something which is not immediately obvious. The only way you can talk about such a problem is to talk about all the ways to solve the problem, and for this you need to utilize real numbers. Trying to solve such problems using traditional algorithms won't work because one needs things like calculus and continuity. To solve this kind of problem, the topological complexity of the method of solving the problem must grow at a certain rate.

My colleagues, Lenore Blum and Mike Shub, and I have extrapolated from an ordinary computer the idea of a real number universal machine which will contain all ordinary machines as a special case. Such a universal machine idealizes the actual computer which can solve all kinds of problems. Development in this theoretical direction is very firmly based on the idea of going from the discrete to the continuous as seen in Newton's *Principia*.

Newton's passage in *Principia* from the corpuscular physical world to his fundamental differential equations of motion set the stage for centuries of developments in science. Three hundred years after *Principia*, the computer revolution is transforming society. We can wonder if the passage from the discrete picture of the computer to a continuous model will deepen our understanding of the laws of computation.

WORKS CITED

Aho, A., J. Hopcraft and J. Ullman. *The Design and Analysis of Computer Algorithms*. Reading, MA: Addison-Wesley, 1979.

Hodges, Andrew. *Alan Turing: The Enigma*. London: Burnett, 1983.

Kuhn, Thomas S. *The Copernican Revolution*. Cambridge: Harvard UP, 1957.

Von Neumann, John. *Collected Works*. 6 vols. New York: Pergamon Press, 1961-63.

Newton's Dream

STEVEN WEINBERG

This publication arises out of a symposium organized to celebrate a great book published 300 years ago, the *Principia* of Isaac Newton. In this book Newton outlined a new theory of motion and a new theory of gravity, and succeeded thereby in explaining not only the apparent motions of bodies in the solar system, but terrestrial phenomena like tides and falling fruits as well. In other work Newton developed the mathematics of the calculus.[1] Newton also performed fundamental experiments in the theory of optics and wrote books about biblical chronology. Yet with all these accomplishments Newton can be said to have contributed to our species one great thing that transcends all his other specific scientific achievements. The title of my paper expresses it – Newton's Dream.

Newton's dream, as I see it, is to understand all of nature, in the way that he was able to understand the solar system, through principles of physics that could be expressed mathematically. That would lead through the operation of mathematical reasoning to predictions which should in principle be capable of accounting for everything. I do not know of an appropriate place in the corpus of Newton's writings to look for a statement of this programme. Newton scholars share with me the feeling that Newton had this aim, but the closest to an explicit statement of it I have found is in the preface to the first edition of the *Principia*, written 301 years ago, in 1686: "I wish we could derive the rest of the phenomena of nature [that is, the phenomena which are not covered in the *Principia*] by the same kind of reasoning as for mechanical principles. For I am induced by many reasons to suspect that they may all depend on certain forces." He wanted to go on beyond the *Principia* and explain everything.

For many years after Newton this ambition took the form of atomism. Clearly, it was a major challenge to explain the behaviour of ordinary materials not in the solar system where the forces are simple – just the long range force of gravity – but here on Earth where things are messy, where axle wheels get stuck in the mud, soup boils on the stove, and an apple left on a windowsill rots. He wanted to explain all of these phenomena by forces like gravity acting on the elementary particles out of which matter is composed.

There is a nice statement of this at the end of his second great book, *The Optics*, first published in 1704.[2] At the end of *The Optics* Newton begins to philosophize about the future of science: "All these things being considered, it seems probable to me that God in the beginning formed matter as solid, massy, hard, impenetrable, moveable particles of such sizes and figures and with such other properties and in such proportions of space as most conduced to the end for which he formed them"; and he goes on later: "It seems to me further that these particles have not only *Vis Inertiae* [which I gather is translated as kinetic energy] accompanied with such passive laws of motion as naturally result from that force, but also that they are moved by certain active principles such as that of gravity and that which causes fermentation and the cohesion of bodies. These principles I consider not as occult qualities but as general laws of nature."

His picture was that the universe consists of particles, the particles are elementary, they cannot be subdivided, they are eternal, but they act on each other through forces – these forces include the force of gravity, but other forces also. Newton was not so foolish as to think that the processes of life or the flow of fluids or the ordinary properties of everyday matter could all be explained in terms of gravity. He knew there must be other forces. But he hoped that these other forces could be discovered and that then the behaviour of matter could be understood through the action of these forces on the elementary particles of matter.

The tradition of atomism, of course, was very old. It pre-dates Newton by several millenia. It goes back to the town of Abdera in ancient Greece, where Democritus and Leucippus had the same conception. But although many before Newton had the idea of explaining everything in simple terms, from Democritus and Leucippus to Descartes, Newton was the first to show how it would work – to show with the example of the solar system how one could explain the behaviour of bodies mathematically and make predictions which would then agree with experiment. Others, as I said, had these grand hopes. René Descartes, who slightly preceded Newton and from whom Newton learned much, also had a broad, general view of nature. But Descartes somehow or other fell short. He never quite grappled with the task of applying quantitatively his principles to the prediction of phenomena. The Newtonian approach – the Newtonian success – had no predecessors, and it left physicists with the challenge of carrying it further.

Atomism continued to be at the centre of this programme for many scientists. In the early nineteenth century great steps were taken in applying the atomic idea to chemistry. Generally speaking, chemists in the nineteenth century became quite comfortable with atoms: they measured atomic weights and knew that there were molecules of water, for instance, which consisted of three atoms, two of hydrogen and one of oxygen. However, atoms seemed hopelessly remote from direct observation. I would imagine that not only Democritus and Leucippus, but also even Dalton and Avogadro, felt that atoms were as far removed from their experimental direct observation as we now think superstrings are removed from direct study. And, in fact, by the end of the nineteenth century there was to some degree a reaction against atomism, and against Newton's dream. Several German and Austrian physicists, led by Ernst Mach, followed a positivistic line according to which physicists are supposed simply to make measurements. Theory is convenient as a method of organizing the measurements but physicists are not supposed to inquire more deeply into things; in particular they are not supposed to ask about atoms which are not directly measurable.

Another challenge to atomism came from the development of an alternative, the field. Again starting in England with the work of Faraday and continuing with the great synthesis of Maxwell, it became possible to think that perhaps the fundamental ingredients of nature are not atoms at all but fields, extended regions within which there is energy and momentum which perhaps form tight knots that we observe as particles. Maxwell's formulation of the field theory of electricity and magnetism in its way is as much of an achievement as Newton's mathematical theory of the motion of bodies under the influence of gravity, except of course that Maxwell followed in Newton's footsteps. When the electron was discovered in 1897, the same experiments were being done in England and in Germany. In England there was a predisposition, because of the tradition of Newton, Prout, and Dalton, to believe in elementary particles, and when these experiments were performed at the Cavendish Laboratory, by J.J. Thomson in 1897, they were immediately heralded as the discovery of the electron. The same experiments were being done in Karlsruhe by Wilhelm Kaufmann who simply reported that he had observed a cathode ray to bend in a certain way under the influence of electric and magnetic fields. The atomistic idea had its first triumph with the discovery of the electron but the field idea was still very strong, and immediately field theorists went to work trying to make models of the electron. The early twentieth century saw a number of extremely elaborate models by Poincaré, Abraham, and others in which the electron was supposed to be just a little bundle of field energy. It was the particular genius of Einstein in 1905 to realize that the time was not yet right for a model of the electron, that in fact one should try to understand the motion of electrons and the behaviour of light in terms of symmetry principles, particularly his famous principle of relativity, and put off for the future the question of the nature of the electron.

I have been discussing Newton's followers only in the context of physics, although in the nineteenth century great steps were also made to realize Newton's dream outside of physics. Perhaps even more important for the intellectual evolution of western civilization were the developments in biology, above all the realization by Darwin and Alfred Wallace that life, with all its apparent purpose and its apparent fashioning for use, can develop through a more or less random series of breedings and eatings. This fact together with the discovery that organic chemicals could be synthesized from inorganic chemicals, led to the realization that life is not something apart from the same general world of phenomena which, following Newton's dream, may someday be explained. As Newton said: "They may all depend on certain forces."

These ideas came together and began to make sense with the advent of quantum mechanics in the 1920s. With that, we understood for the first time what atoms are and how the forces that act within atoms, specifically the electromagnetic force, produce all the rich variety of chemical behaviour which, over the course of billions of years, produced the phenomenon of life. And although Newton started the dream, and many great contributions were made to it in the ensuing centuries, the beginning of the realization of Newton's dream is with the advent of quantum mechanics and its explanation of the nature of ordinary matter.

By the mid-1920s one could say that in some sense all natural phenomena could be explained, at least in principle, in terms of the particles that make up ordinary matter. These particles were at that time conceived to be electrons, which are the particles in the outer parts of atoms, discovered by Thomson (and if he had wanted to claim credit for it, Kaufmann); photons, the particles which in large numbers make up a ray of light; and atomic nuclei, which were then mysterious, positively charged, masses at the centres of atoms, that hold the electrons in their orbits by the force of electrical attraction, and which contain by far the greatest part of the mass of ordinary atoms. And then one other ingredient in the universe, not understood at all, seeming to have nothing to do with phenomena at the atomic scale, the force of gravity.

The physicists of the 1920s had what seemed a fairly simple world. They didn't understand anything about nuclei. They knew about protons, which are the nuclei of the simplest atom, the atom of hydrogen, and which are particles with just one positive unit of charge. And they knew that every once in a while nuclei spit out electrons. It seemed obvious to them that nuclei consist of protons and electrons. There were terrible problems, theoretical problems, in trying to make a picture of a nucleus consisting of protons and electrons holding together, but it was hoped that somehow all would be understood in the future. I suppose the physicists of the 1920s would have said the universe consists of electrons, protons, and also gravity, which seems to have nothing to do with the other particles. There is a famous remark made just after Paul Dirac formulated his relativistic theory of the electron in 1928. Someone, whose name I cannot recall, said: "In two more years, we will have the proton [meaning we will have the relativistic quantum theory of the proton] and then we will know everything." Well, that didn't happen.

The 1930s saw a period of intense study of atomic nuclei. It was discovered that nuclei consist not of protons and electrons but of protons and other particles called neutrons. After World War II, when a new generation of accelerators came on line, it began to be possible to produce entirely new species of particles, particles that seemed like siblings of the electron or the proton or the neutron, but were heavier. And because they were heavier (so that there was energy available) they could easily decay into the lighter particles and did so. They lived very short lives and were produced in laboratories at Berkeley and Brookhaven. The years after the mid-1950s saw an attempt to come to grips with the expanding population of species of particles, and by the mid-1970s it had fallen into place with the formulation of what is now called the "standard model."[3] The formulation of the standard model of weak, electromagnetic, and strong interactions was finished by the mid-1970s, and by then there were enough experiments that one could actually rely on it. In the standard model the list of fundamental ingredients of the universe still includes the electron, but also its siblings, particles which are quite like the electron and which can turn into electrons under the influence of the weak force, particles called neutrinos, and muons, and tauons. The proton and the neutron no longer are seen to be basic. They are seen as mere bound states, something like an atom, or a molecule, or a blackboard eraser, made up of very elementary particles. The elementary particles of which the protons and the neutrons are supposed to be composed are known as quarks. A number of different families of quarks appeared which had to be named. The first two were called Up and Down, because one had a positive and the other a negative charge. The quarks that followed were called Strange, because they are present in particles whose discovery had come as a surprise, and then Charm for no clear reason, and then Top and Bottom. (I co-authored a very speculative paper during this period in which there happened to appear seven types of quarks; we decided to call them Gluttony, Envy, Sloth, and so on.)

The electron and other similar particles collectively are called leptons. There is apparently a parallelism between quarks and leptons. There are exactly the same number of flavours of quarks and leptons and we think we know why that is true. The particles transmit the forces. We have known about the photon ever since Einstein proposed it in 1905, but we now know the photon has much much heavier siblings that transmit the weak nuclear force. The W particle and the Z particle, as they are called, were discovered only in 1983 at CERN, the elementary-physics laboratory in Geneva. There is another similar class of particles that carry the strong nuclear force. They are called gluons because they glue the quarks together inside the proton and neutron and, as a by-product, also glue the proton and neutron together inside the atomic nucleus.

The photon is a sibling of the W and the Z, but they have vastly different masses; even though there is a family relation among them they seem very different. We now understand this in terms of a phenomenon called spontaneous symmetry-breaking. This refers to the fact that a set of mathematical equations can have a high degree of symmetry (that is, can retain the same form when the different variables are transformed into each other) and yet the solutions of those equations may not share that symmetry. On the level of the underlying equations that govern the particles, the photon, the W, and the Z all appear perfectly symmetrical. There are transformations that change a photon into a W and leave the form of the equations unchanged. Yet the solutions of the equations, which are the particles themselves, don't exhibit that symmetry. They are very different. Once one sees through the breaking of the symmetry, one can easily understand the process.

We do not yet know, however, the mechanism by which the symmetries are broken. A similar phenomenon occurs in superconductivity (in fact that is where it was discovered first). Superconductivity is on everyone's minds these days, but elementary particle physicists in the late 1950s had already learned much about it from their brethren in solid state physics. In the theory of superconductivity, the symmetry which is broken is ordinary electromagnetic gauge invariance, or to put it more simply, conservation of charge. This symmetry is broken by forces which are produced by the exchange of sound waves (or, in quantum sense, phonons) between the electrons in the superconductor. The question is, what plays the role of the sound waves in elementary particle physics? What are the forces that induce the instability which produces the breakdown of the symmetry? We don't know. There are some simple pictures of these forces. The simplest involve certain particles which have come to be known as Higgs particles because they first appeared in a mathematical model invented by Peter Higgs of Edinburgh. This is the part of the standard model that we are hoping will be clarified by experiments at the great new accelerator which we hope will be built in the next six or seven years, the superconducting supercollider (SSC).

It is clear that the standard model is not the final realization of Newton's dream. Even when we get straight about these Higgs bosons there will still be too many arbitrary features in the standard model. Even if there were no particles waiting to be discovered besides those which we know are theoretically necessary – specifically one more quark, the top quark, and the Higgs boson – then even so, the standard model has in it seventeen numerical quantities like the charge of the electron, the mass of the electron, the masses of the quarks, and so on. We learned why they have the values they have from experiment, but we don't know why nature chooses those values. Any theory that has seventeen free parameters is too imprecise to be satisfactory.

There has to be something else. The standard model leaves out gravity. Gravity is still doing its job in the solar system, the way it was in the 1920s, having no observable effect at the level of atoms and molecules, and being absolutely inaccessible to experimental study except on macroscopic scales, where we have enormous numbers of particles adding up their gravitational fields. We need a theory that goes beyond the standard model because we want to explain those seventeen parameters. More than that, we want to explain the standard model itself. We want to explain why there have to be all these flavours of quarks and leptons. And of course we want to bring gravity into the picture.

Many theorists feel that the next step will take the form of superstring theories, in which the basic ingredients of the universe are seen not to be particles, not to be fields, but instead to be something like rubber bands, little strings that go zipping around, vibrating in many normal modes. The normal modes of vibration are what we see as different species of particles, but they are all one kind of string. This kind of theory has a long way to go before it is able to explain things like the values of the seventeen parameters of the standard model. I have taken as landmarks in this discussion the 1920s synthesis in terms of the quantum mechanics of electrons and protons, and the 1970s synthesis of the standard model, so by an arithmetical progression, we really shouldn't expect any new breakthroughs until 2020. It has been said that the reason we are having so much trouble doing superstring physics is that it is twenty-first-century physics which we accidentally stumbled on thirty years too early.

The shape of the final realization of Newton's dream is very far from clear. We believe, though, that it will have to include quantum mechanics. Since the development of quantum mechanics in the 1920s there has been not the slightest suggestion that any physical phenomena require anything beyond quantum mechanics. And it is always the quantum mechanics of 1925 to 1926, the quantum mechanics we learn in our first course. The quantum mechanics worked out by Schrödinger and Heisenberg and Born and Jordan and Pauli in 1925 to 1926 is the same quantum mechanics we apply to quantum fields in the standard model, and it is the same as we apply to superstrings today. Quantum mechanics seems to be a permanent part of our physical understanding. Lately

I have been trying to understand why that is, by asking what kind of generalization of quantum mechanics could possibly be logically consistent. Is there any way, for example, of taking the linear equations of quantum mechanics and introducing nonlinearities? I can tell you it is very hard. It is very hard to think of any way of tampering with the rules of quantum mechanics without having logic fly out of the window, for example, without introducing negative probabilities or probabilities that don't add up to one. So I find it difficult to imagine that the future synthesis of everything will not be in the language of quantum mechanics. Quantum mechanics is a grammar in terms of which all physics must be expressed, but it does not itself tell us anything. That's one of the reasons it is so hard to test, because by itself it says nothing. The other ingredient that seems to be needed to add to quantum mechanics to complete a picture of the universe is the symmetries. Now that may surprise you because in many areas of science symmetries are somewhat incidental. Biologists, for instance, know that a human being has roughly a symmetry that the mathematician calls z_2, the interchange of right and left, but that is certainly not the most interesting thing about human beings. In particle physics, it seems that symmetries are the most interesting things about elementary particles. Think of what it takes to describe an elementary particle. One describes an elementary particle by giving its momentum and its energy and its charge and a few other things. Every one of these is simply a number that describes how the wave function or the state vector of the particle changes when symmetry transformations are performed. Aside from their symmetry transformation properties, all particles are the same. There may be nothing to nature but quantum mechanics, which is the stage for physical phenomena, and the symmetry principles, which are the actors. But that's looking far ahead. We do not know that there aren't entirely new ideas which will have to be invented. The progress in superstring theory has not been as exciting in 1987 as it was in 1984 or 1985. And it is beginning to look like some very different new ideas are needed to make further progress there.

Now that I have given you a complete history of science in the last three hundred years, I would like, if I may, to try to draw some lessons from this experience. Those of us whose lives are governed by our own little versions of Newton's dream are trying to accomplish something very special. Our work, although we are delighted if it has some utility, is not particularly directed toward utility. Nor have we chosen the problems we are working on because they are fun or mathematically interesting. Sometimes we are accused of that and in fact, sometimes, as a kind of self-protection and to avoid the accusation of taking ourselves too seriously, we claim that we do the work we do just for fun. But that is not all there is to it. We, meaning the community of elementary particle physicists and those in the related disciplines of cosmology and astrophysics, have a historical goal in mind. The goal is the formulation of a few simple principles that explain why everything is the way it is. This is Newton's dream and it is our dream.

This sense of historical direction makes our behaviour somewhat different from that of scientists in general. For one thing, the importance of phenomena in everyday life is, for us, a very bad guide to their importance in the final answer. The fact that the electron is ubiquitous in ordinary matter is simply because the electron is lighter than the muon. The muon is two hundred times heavier than the electron so there is lots of energy available for the muon to decay into the electron and a couple of neutrinos, which it does with a lifetime of a microsecond or so. We do not know about muons in our everyday life. But as far as we know, muons play just as fundamental a role (which may or may not be very fundamental) as electrons in the ultimate scheme of things. The fact that electrons were discovered first and are far more important to matter throughout the universe is of secondary importance. This point often comes up because, although muons are present in cosmic rays, by and large these particles are particles that we have to produce artificially in the laboratory. One may imagine that we are like butterfly collectors studying butterflies that we have created ourselves in our laboratory and which do not exist in the real world. We are not very interested in our butterflies. We are not particularly interested in our electrons or our muons. We are interested in the final principles that we hope we will learn about by studying these particles. So the first lesson is that the ordinary world is not a very good guide to what is important. I suppose this lesson could have been learned before. Copernicus, you remember, thought that the orbits of planets had to be perfect circles because something as important as a planet would have to move on a perfect curve, which is a circle. Today we don't look at planets as being very special. Planets are just bodies that happen to have formed in the history of the solar system and whose orbits have various eccentricities, but there is no particular reason why their orbits have to be circles.

This brings me to the second lesson. It is that if we are talking about very fundamental phenomena, then ideas of beauty are important in a way that they wouldn't be if we were talking about mere accidents. Planetary orbits don't have to be circles because planets are not very important on some fundamental level. On the other hand, when we formulate the equations of quantum field theories or string theories we demand a great deal of mathematical elegance because we believe that the mathematical elegance that must exist at the root of things in nature has to be mirrored at the level we are working. If we thought the particles and fields we were working on were mere accidents that happened to be important to human beings, but were not themselves special, then the use of beauty as a criterion in formulating our theories would not be so useful.

Finally, the kind of beauty for which we look is special. Beauty, of course, is a general and broad and vague word. We find many things beautiful. The human face is beautiful, a grand opera is beautiful, a piano sonata is beautiful. The kind of beauty we are looking for is more like the beauty of a piano sonata than that of a grand opera, in the specific sense that the theories we find beautiful are theories which give us a sense that nothing could be changed. Just as, listening to a piano sonata, we feel that one note must follow from the preceding note - and it could not have been any other note - in the theories we are trying to formulate, we are looking for a sense of uniqueness, for a sense that when we understand the final answer, we will see that it could not have been any other way. My colleague, John Wheeler, has formulated this as the prediction that when we finally learn the ultimate laws of nature we will wonder why they were not obvious from the beginning.

That may very well be true. If it is, I suspect it will be because by the time we learn the ultimate laws of nature, we will have been changed so much by the learning process that it will become difficult to imagine that the truth could be anything else but superstring theory - or whatever it turns out to be.

So far, I have been speaking in a style which is sometimes called physics imperialism. That is, the physicist provides a set of laws of nature which explain everything else and all the other sciences appear to be offshoots of physics. I want, at least in part, to disavow this. I do believe there is a sense in which everything is explained by the laws of nature and the laws of nature are what physicists are trying to discover. But the explanation is an explanation in principle of a sort which doesn't in any way threaten the autonomy of the other sciences. We see this even within physics itself. The study of statistical mechanics, the behaviour of large numbers of particles, and its applications in studying matter in general, like condensed matter, crystals, and liquids, is a separate science because when you deal with very large numbers of particles, new phenomena emerge. To take an example I have used elsewhere, even if you tried the reductionist approach and plotted out the motion of each molecule in a glass of water using equations of molecular physics to follow how each molecule went, nowhere in the mountain of computer tape you produced would you find the things that interested you about the water, things like turbulence, or temperature, or entropy. Each science deals with nature on its own terms because each science finds something else in nature that is interesting. Nevertheless, there is a sense that the principles of statistical mechanics are what they are because of the properties of the particles out of which bodies are composed. Statistical mechanics does not have principles that stand alone and cannot be deduced from a deeper level. As it happens, that deeper level is always a more microscopic level. If we ask any question about nature – why the sky is blue or the grass is green – and keep asking why, why, why, we will get a series of answers which always takes us down to the level of the very small.

There is a sense in which the kind of thing that elementary particle physicists study is especially fundamental, but in no way does it threaten the separate existence or the special importance of other sciences. Because we physicists think we are moving toward the final answer, the work we are doing is not necessarily more worthy of support than the work of other scientists. This has come up in Britain with the debate over further participation in CERN, and recently in the US in the debate about whether or not to spend $4.4 billion for the SSC accelerator. It is argued by the opponents of CERN in Britain and the opponents of the SSC in the US that elementary particle physics is no more fundamental than other areas of science; it is less likely to yield results of direct practical importance, and therefore it should not take such a large share of the public funds of these countries. These are very difficult questions, and I would have a hard time myself deciding which areas of science deserve what part of a research budget. It may not be the issue, however, because there is evidence that spending on large research projects like the SSC actually helps to buoy up spending on research in general. But I don't want to get into that argument because it's a matter in which I don't have any expertise. I want to make only one point, but that point I want to make very strongly, that although all of the sciences have credentials to justify their support, credentials that may take the form of practical utility, or impact on neighbouring sciences, or intellectual challenge, there is one particular credential that elementary particle physics has which is not necessarily more important than the others but is worthy of respect. And that is that we are trying to get to the roots of the chains of explanation, that we are trying to grapple with nature at the most fundamental level that it is given to human beings to address in history. In other words, that it is we who are trying to realize Newton's dream.

NOTES

1 He was apparently not entirely comfortable with calculus, and played it down in the *Principia*, but it is the method by which today we carry out Newton's calculations. It is very hard for a modern physicist to read the *Principia* because its style is so geometric.
2 I quote here from a later edition.
3 Standard model is a wonderful term. I have used it in a book on gravitation and cosmology to describe the Big Bang theory of the universe, and I understand biologists also have a standard model. It is a marvelous expression which means that which we all take as our working hypothesis, that which it is respectable to write scientific papers about, but which by no means we are absolutely committed to believe.

Symmetry in Art and Nature

DENYS WILKINSON

When we seek to describe the way in which the world is built, that is to say, grandly speaking, the structure of the universe, we inevitably bring to that attempted understanding the prejudices that derive from our own nature and everyday experiences. But our human nature ranges over matters such as love, compassion, beauty, greed, worship, hatred, and terror that, we presume without certainty, have little to do with the structure of the material world. On the other hand, our own everyday experiences cover only a tiny fraction of the scales in space and time that we are led to contemplate in our discussions of the physical world.

We must therefore recognize that, in one sense, the input to our attempt to understand the universe through our humanity is vastly over-rich. So much of our own essential human nature seems to be irrelevant for our task of building the material universe, while, in a complementary sense, our own personal experiences of the scales of space and time are undesirably minute and inadequate. We are forced to make gigantic leaps of presumption beyond what we can directly "know." We must make those leaps because, by definition, we do not know how to do anything else. The question is, how far can we go?

Scales of Space and Time

We should pause, briefly, to quantify that second consideration, namely the unimaginably small window, in space and time, through which we, as humans, glimpse the world of nature. Begin with the fact that the smallest object that we can see is about 10^{-4} cm across[1] while the largest of which we can claim any kind of reasonable personal experience is about 100 km. But the smallest distance that we have so far been compelled by our experiments to consider is the present upper limit to the size of the electron of about 10^{-16} cm. This is about 10^{12} times smaller than we can see, while the smallest distance that we contemplate through our current theories is, as we shall see later, the Planck length of 10^{-33} cm, i.e., 10^{29} times smaller than we can see. On the large side we must discuss the 10^{23} km or so that we infer as the distance from us of the remotest observed galaxies, i.e., 10^{21} times larger than our personal experience.

And similarly with time, the shortest interval that we might reasonably claim to appeal directly to our own experience is about 1/100 of a second and the longest about 100 years. The shortest time of our present inferential experience is about 10^{-26} seconds, namely the time taken for light to traverse that "shortest distance" of 10^{-16} cm that I mentioned earlier, while the shortest time entering our current theories is the Planck time of 10^{-43} seconds; these are times shorter, respectively, by factors of 10^{24} and 10^{41} than our personally-perceived shortest interval. On the long side, we believe that our universe has an age of about 10^{10} years, a factor of 10^{8} beyond our own grasp; but in terms of our physical theories we happily discuss the lifetime of the proton of more than 10^{32} years and the lifetimes of massive black holes of 10^{100} years and more.

These factors by which the great and the small, the long and the short, of our discussion of the natural world are remote from our personal experience are totally beyond our operational comprehension. So in these two senses we are not very well matched into the natural world that we seek to describe and understand. We are too rich in the qualitative range of our human experiences and too poor in the range of our physical experiences. In these circumstances we must admit that our description of the world must have a certain anthropomorphic quality as we see it through our human prejudices and make those unimaginable extrapolations into unknown space and unknown time.

We are going to examine here just one of the prejudices that we bring to our description of the physical natural world and which is most obviously expressed through art, namely symmetry. Much of art, particularly architecture, is in all cultures highly symmetrical: think only of temples, castles, and palaces, edifices of all kinds. For some reason we feel that symmetry is right: perhaps it is because we are, superficially, ourselves symmetrical. We bring to our description of nature the prejudice that symmetry is good and that we shall understand how the world works if we can recognize the symmetries to which it conforms.

Symmetries and Conservation Laws

There are many forms of symmetry apart from the mirror symmetry of ourselves and of temples. The stylized letter S and artistic motifs derived from it do not have mirror symmetry but they have a rotational symmetry in that they are unchanged by a rotation of 180° about their mid-point. A long avenue of trees has spatial translational symmetry in that its aspect is unchanged by moving all the trees over by one space.

In fact, symmetries are no more than statements as to the operations that *have no effect upon* the systems that we consider. We feel that if we can make such statements with appropriate generality we have discovered a Law of nature, i.e., we have learned something about how the world works. Many elementary and familiar principles can be recast in the language of symmetry. The law of the conservation of linear momentum is the same as the statement that Nature possesses spatial translational symmetry: if you carry out an experiment with the apparatus at one place on the laboratory bench you will get the same answer if you repeat the experiment at another place along the bench: the regular avenue of trees that we feel "looks right." Similarly the law of the conservation of angular momentum is the same as rotational symmetry: it will make no difference if you turn the apparatus through an arbitrary angle and do the experiment again. And the law of the conservation of energy is the same as translational symmetry in time: if you do the experiment again tomorrow you will get the same answer as today. Symmetries, as generalized to the whole world of physical nature, are then just those operations of all kinds that leave the system to which they are applied operationally unchanged.

It is not too much to say that our attempt to understand the workings of nature is the attempt to discover those symmetries to which it conforms. Of course, the symmetries do not, of themselves, tell us what will actually happen because that depends upon the details of the system to which they are applied: the rules of football do not enable you to predict the result of the match. The symmetries of nature are the rules of the game, they do not say what will be the details of play. So far we have reasoned as though symmetries were exact as, perhaps, some of them may be although we cannot know with certainty. All natural laws are empirical laws and may have only a limited range of validity. As we shall see later some are obeyed to a very high degree of precision in one context and are completely violated in another.

We have here a further comparison between art and nature. We start out by considering a perfect symmetry but depart from it when there is good reason; in the jargon, the symmetry is broken. As we shall see, such broken symmetries are richer and more revealing of nature's interconnections, the Grand Plan, than are simple perfect symmetries.

Broken Symmetry in Art

I mentioned the symmetry of temples. Some are almost exactly symmetrical as, for example, the great cathedral of Notre-Dame de Paris, the west façade of which, begun in the late twelfth century, is an almost perfect mirror image of itself. (Although it is not exact: for example the north portal – the Virgin's – differs slightly from the south portal – St Anne's – following the medieval feeling that exact symmetry entrained monotony and should be slightly broken even

Figure 1. Notre-Dame de Paris. [From Marcel Aubert, Notre-Dame de Paris: Architecture et Sculpture. *Paris: A. Morancé, 1928.]*

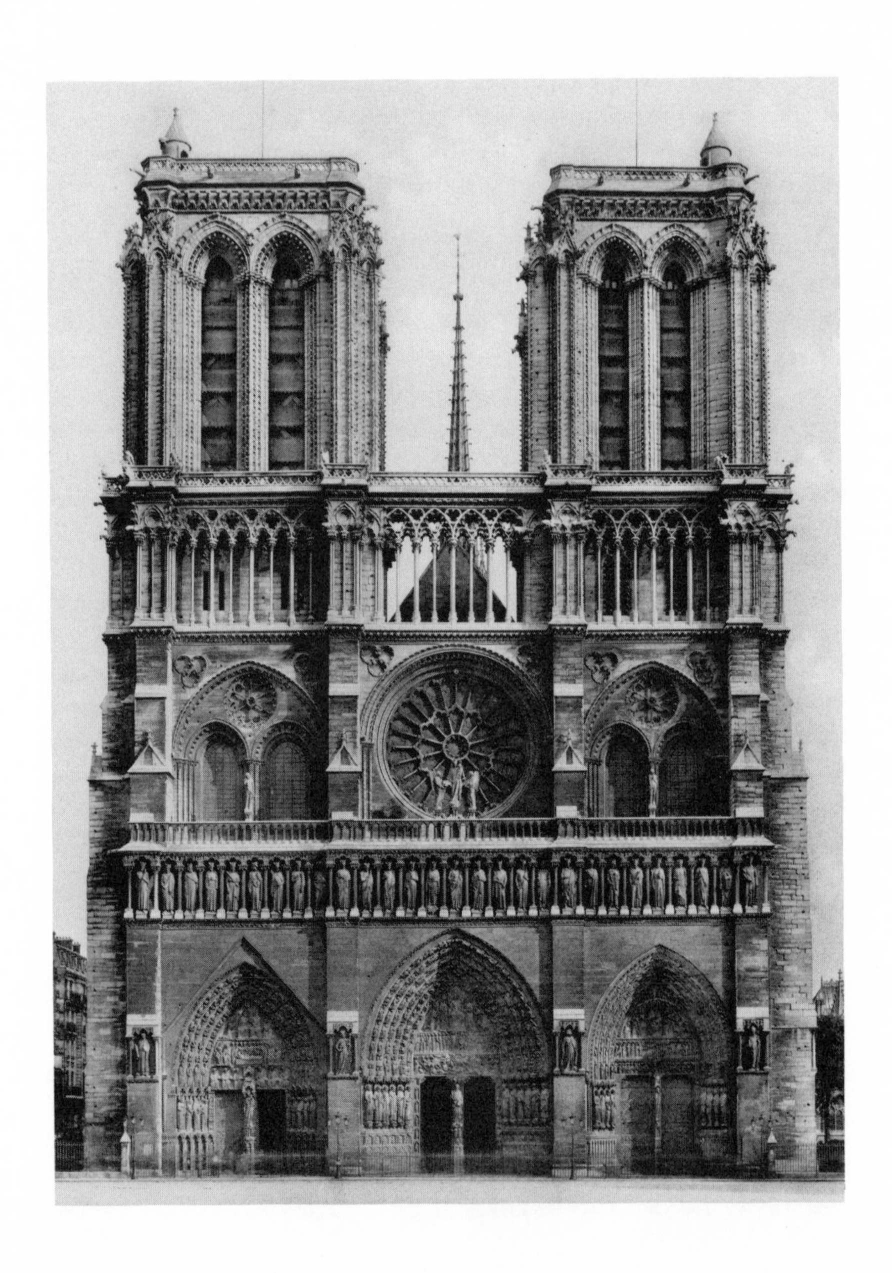

when that was not otherwise "necessary.") (Fig. 1)

As in the Paris cathedral, the lower stages of the façade of Notre-Dame de Chartres (Fig. 2) are very largely symmetrical. But in the spires the symmetry is dramatically broken: that to the south is a clean, austere, beautifully proportional octagonal pyramid rising to 105 metres; that to the north is a fantastical, flamboyant, multi-crocketed confection outsoaring its astonished and ashamed companion by fully ten metres. Can we understand the reason for this gross breaking of the early Gothic symmetry? We may (only half-fancifully) imagine that after completion of the southern spire the money ran out so that the northern spire could not be added until the beginning of the sixteenth century, three hundred years later. By this time architectural fashion, even more tyrannical than that of *haute couture*, had changed to the degree that it would have been unthinkable to provide a purely symmetrical echo to the *clocher vieux* of despised ancient days. But how much richer is, in fact, the broken symmetry in its extra dimensions of art, history, and economics than the perfect symmetry that might have been! And so we shall see it to be with nature also.

Before we leave this first example of broken symmetry I must draw attention to an important aspect of it, the arbitrariness of the sense of the symmetry-breaking. At Chartres the symmetry could have been broken to just the same effect the other way round: the *clocher vieux* to the north and the *clocher neuf* to the south. For some unimportant reason the symmetry broke the one way round but from the point of view of its impact upon us it could as well have been the other way.

Another striking, but less familiar, example of symmetry-breaking is the famous painted limestone bust of Queen Nefertiti from Tell el-Amarna now in the Berlin Museum (Fig. 3). She has only one eye. This is so shocking in such an exquisite and sensitive portrait that the reproductions that sell in their thousands all over the world have restored the symmetry. Nefertiti's one eye is inlaid with crystal set in a black paste which creates the iris; the other socket, geometrically and structurally symmetrical with the first, is empty. There is no evidence from inscriptions or other attributed portraits that Nefertiti was blind in one eye. Why is the symmetry broken? Did one eye fall out? Did Thutmes, the sculptor of 1360 BC, fear a Pygmalion-like love for a too-perfect beauty? It is likely that Nefertiti is only a maquette: perhaps the commission was cancelled before the model was finished.

For whatever reason, the symmetry is broken, one eye is missing. But which? Again we see that the sense of the breaking of the symmetry is unimportant as well as the reason for it; both are irrelevant to its effect. Again we see that the breaking of the symmetry spoils perfection in a way that enhances our appreciation.

Figure 2. Notre-Dame de Chartres. West façade. [Photograph by James Austin.]

Figure 3. Bust of Nefertiti [Collection Ägyptisches Museum der Staatliche Museum, Berlin.]

Symmetry-breaking in art is also seen in a seventh century Irish illuminated manuscript of St Matthew from the Book of Durrow, now in the library of Trinity College, Dublin. St Matthew is remarkably symmetrical. (Fig. 4) But look at his feet; why do they so strikingly break the full-frontal symmetry? Consider the alternatives: if the feet turned symmetrically outwards Matthew would present a ludicrous Chaplinesque appearance; if they turned symmetrically inwards he would look even sillier; if they were both straight forward he would seem to be standing on little pegs. Here the breaking of the symmetry is almost forced by the symmetry itself and the commitment to a certain stylistic representation. But note again the arbitrariness of the sense of the breaking: to have his feet pointing to the right would have served St Matthew just as well. (In which case we might have understood the sense of the symmetry-breaking by a reading of Ecclesiastes 10.11: "The heart of a wise man inclines him towards the right but that of a fool towards the left.")

Broken Symmetry in Animate Nature

When we turn to animate nature we again seem to be dominated by superficial symmetry, although many creatures break their symmetry. Crossbills have bills that do not meet symmetrically but rather cross over at the tip. This symmetry-breaking is explained by the fact that crossbills earn their living by prying open pine and other cones to eat the nuts. The muscles that close mouths can exert more closing force than those that open them can exert opening force, so its bill design enables the crossbill to apply more opening force on the cone by closing its mouth than it could by opening its mouth if its bill were not crossed. (Young birds are fed by their parents and so do not need to open their own cones. Their bills do not cross until one or two weeks after fledging.) In this case the reason for the symmetry-breaking is understood. But do the bills cross clockwise or anti-clockwise? Surely it can make no difference? It does not: both senses of crossing, of symmetry-breaking, are found.

Owls depend for their location of prey quite largely upon sound. Sound location itself depends on the time difference between the signals received by the two ears and on the relative intensity of the two signals. A symmetrical arrangement of ears leads to ambiguity in the inference as to the location of the source of the sound. Thus signals arriving simultaneously and with equal intensity at two identical symmetrically-placed ears would merely inform the owl that the prey was located in the vertical plane bisecting the owl. If the ears are identical but one points up and the other points down the relative intensities of the two signals will tell the owl in which direction on that vertical plane the prey is to be found. Non-identical ears can be even more informative. Such asymme-

Figure 4. Symbol of St Matthew [From the Book of Durrow, *in Trinity College Library, Dublin.]*

tries, of which this is a simplified example, are found in the ears of many species of owl.

We have seen that in art and in animate nature symmetry is often spontaneously broken in response to some sort of environmental factor of need and that the sense of the breaking is then random or arbitrary. Symmetry-breaking has also a long history of philosophical fascination. The best-known example is that of Buridan's Ass. Jean Buridan, who lived from about 1295 to some time after 1366, was rector of the University of Paris, in his science a precursor of Galileo and Descartes and a defender of causality and determinism.[2] Symmetry enters when Buridan's Ass, hungry, finds himself exactly midway between two equally attractive bundles of hay and starves to death because there is nothing in his situation to enable him to make up his mind as to which bundle of hay to go to – victim to the ancient Latin tag *liberum arbitrium indifferentiae*. (In point of fact, Buridan's writings do not contain this parable. It was invented by his opponents to discredit him.) Of course, the perfect hypothetical symmetry of the dilemma of Buridan's Ass cannot be achieved and even if it could, it could not be maintained; the situation is unstable so some perturbation, no matter how slight, would incline the ass to the one bundle of hay or to the other, the symmetry would be broken and the ass would live. But notice, as with our earlier examples from art and nature, that the symmetry might break either way, left or right.

To a scientist this is the problem of the ball balanced on the sugar-loaf crown of the sombrero: symmetry says that the ball will stay on the crown but an indefinitely-small disturbance causes it to roll down into the trough of the brim. But we cannot say where around the brim – here the symmetry can be broken in an infinite number of ways.

The Physical World and the Granularity of Physical Nature

After this somewhat discursive foray into symmetry-breaking in a range of contexts we shall turn back to the physical world to see how these same ideas lie behind our present thinking about natural science and the cosmos itself.

Ordinary matter, such as ourselves, is made out of *atoms* which are small in relation to ourselves – their dimensions are a few times 10^{-8} cm. At the centre of the atom is the nucleus, of dimensions a few times 10^{-13} cm, surrounded by negatively charged electrons, whose locations define the geometrical extent of the atom, but which are individually no more than 10^{-16} cm across. So the atom, like the solar system, is chiefly empty space. Even the central nucleus of the atom is chiefly empty space. Indeed, the atom is even emptier than the solar system. In the solar system the planets occupy about 3×10^{-15} of the volume (and represent about 0.1 per cent of the mass) whereas in an atom the electrons occupy less than 10^{-23} of the volume (and represent only a few times 0.01 per cent of the mass).

The next level of granularity below that of the atom is that of the *nucleus*. The nucleus is made out of positively charged protons, and in light nuclei such as those of oxygen and calcium about as many neutrons, which are electrically neutral but otherwise almost identical to protons. In heavy nuclei such as those of lead there are about half again as many neutrons as protons. Note that the positive electrical charge of each proton is numerically equal to the negative charge of each electron to better than one part in 10^{21}. To this astonishing equality, for which there is no obvious reason, I shall return later. Unlike the atom, the nucleus does not appear, at first sight, to be chiefly empty space. However, when we go to the next level of granularity, below the neutron and proton, we see that the packed nature of the nucleus is illusory.

We have known for some twenty years that the neutron and proton are not elementary particles but are each made of three constituent entities called *quarks*. The quarks appear to be confined within the neutrons and protons by forces that increase as the separation between the quarks is increased, much as the tension in a rubber band increases the more we stretch it. We have not been able, in consequence, to remove quarks from neutrons and protons to study them as separate individual entities. But we have been able, through a wide range of experiments, to infer the properties and behaviour of the quarks within the proton even though we cannot examine them separately. Recent experiments have revealed that the quarks are about one hundred times lighter than the neutron and proton that comprise them. The rest of the mass corresponds to the energy of the quarks' interactions and motions. The quarks, like the electrons, appear to our experiments as points (certainly less than 10^{-16} cm across) and so occupy no more than, at most, about 10^{-9} of the volume of the neutron and proton. The neutron and proton, and therefore the nucleus, are consequentially, like the atom, mainly empty space.

Here, at least for the moment, the hierarchy of granularities ends; we have so far encountered no sign of substructure within the electron or within the quarks although we should be rash to assert that the end of the road has been reached and that electrons and quarks are indeed irreducibly elementary particles.

Quark Flavours and Colours

Before continuing we must pause for some filling-out of the picture. We should know that the three quarks within the neutron and the proton are of two different types (called *flavours*); up (u) and down (d). The u-quark carries an electrical charge of $+2/3$ if we call the charge of the electron -1 (and, therefore, that of the proton +1 and that of the neutron 0); the charge of the *d*-quark is −⅓ so that the three quarks of the proton are written *uud* and those of the neutron are *udd*. Furthermore, each flavour of quark, *u* and *d*, comes in three different and distinguishable guises, the so-called *colours*: red (R), green (G), and blue (B) chosen for the primary colours of human vision that together make up white. Physicists thus describe quarks as u_R, u_G, and u_B and similarly d_R, d_G, and d_B. They believe that the mysterious confinement of the quarks within the neutron and proton can be illustrated by the remark that a "coloured" particle, a single elementary quark, or a coloured combination of quarks, cannot exist as a free entity and that the only particles having separate existence in the free state (such as protons and neutrons) must be "colourless," i.e., "white" in the sense of containing one red, one green, and one blue quark.

Isospin Symmetry

We may note in passing that we have just revealed a symmetry. We have seen that the neutron and proton are very much the same thing, apart from their electric charges. Physicists have also discovered another symmetry operation, namely, changing the charge for particles such as the proton has no essential effect on the system concerned. This is called *isospin symmetry*, the independence of the structure and interactions of the particles when their electrical charge changes.

Anti-Particles

At this point I would like to introduce an all-pervading concept of modern physics, the *anti-particle*. The fusion by Paul Dirac in 1928 of Albert Einstein's special theory of relativity with the earlier quantum theory leads to the expectation that certain sorts of particle, such as electrons and quarks, will possess inalienable and unalterable angular momentum of their own, called intrinsic spin. The same relativistic quantum theory leads to the requirement that every particle in nature should possess an associated anti-particle which, crudely speaking, has all the reversible attributes of the particle reversed but is otherwise identical to it. We now come to one more symmetry: nature is invariant under the transformation of particle into anti-particle (designated by physicists by C for charge conjugation). If a physical system can exist so also can that system with all particles changed into their anti-particles in the same states of motion.

If indeed particle and anti-particle have exactly equal and opposite attributes then we must be able to create them in pairs, particle plus anti-particle (by supplying energy equal to or greater than $E = 2mc^2$ where m is the particle/anti-particle mass). This is called pair creation and works according to plan. Conversely, if we bring together a particle and its anti-particle the attributes can cancel out the anti-attributes leaving nothing but the $2mc^2$ of pure energy. This is called annihilation and also works exactly according to plan.

Anti-Quarks, Mesons, and Baryons

We now return to our *u*- and *d*-quarks. To the *u*-quark there must correspond the anti-*u*-quark or $\bar{u}$ (anti-particles are designated by placing a bar over the symbol of the particle) which will have electrical charge $-2/3$ and also the colour anti-R, anti-G or anti-B, viz. $\bar{R}$, $\bar{G}$ or $\bar{B}$, e.g., $u_R \rightarrow \bar{u}_{\bar{R}}$. But now $R + \bar{R}$ is, by definition, colourless since we can make $u_R + \bar{u}_{\bar{R}}$ out of pure energy which is certainly colourless.

This, then, is our second way of making a colourless, and therefore, by our earlier hypothesis, permissible particle. The colourless $u_R\bar{u}_{\bar{R}}$ should exist although it probably will not last very long because the $u_R\bar{u}_{\bar{R}}$ quarks can annihilate into pure energy. However, an object such as $u_R\bar{d}_{\bar{R}}$ cannot annihilate into pure energy because the charge of the $\bar{d}$ is the reverse of that of the *d*, namely $--1/3 = +1/3$, while that of the *u* is $+2/3$ so that the charge of the $u_R\bar{d}_{\bar{R}}$ is $2/3 + 1/3 = 1$. So the $u_R\bar{d}_{\bar{R}}$ will not annihilate itself into pure energy.

Such combinations of quarks and anti-quarks exist in great profusion, representing a wide range of possible such combinations (some 30 are now known). These combinations are called *mesons*. The lightest of the vast array is called the pion (π) with a mass some 270 times greater than that of the electron, 7 times less than that of the proton. These quark rearrangements in threes generate the spectrum of so-called ordinary *baryons*. These are generated experimentally by exciting (i.e., by applying energy to) neutrons and protons. An atom in this situation disposes of its excess energy by emitting its characteristic radiation – light. In the case of particles such as those we are now discussing the characteristic radiation is mesons. Thus inter-relationships and inter-transformations of the rich spectra of baryons and mesons are most satisfactorily understood, quantitatively, in terms of the quark rearrangements. Baryons and mesons are collectively known as *hadrons* and their interactions with one another are called the nuclear interactions; we shall return to these shortly.

Radioactivity: The Weak Interaction

The neutron is slightly heavier than the proton. In free state, i.e., as an isolated particle in the laboratory, it lives for only about 10 minutes before emitting an electron, releasing kinetic energy, and becoming a proton. At the same time as it emits the electron, the neutron also emits a very light, possibly massless, electrically neutral particle, the neutrino (ν) so that the full reaction is $n \rightarrow p + e^{-} + \bar{\nu}$. (Technically the particle involved in this particular reaction is the anti-neutrino.) The neutron decay is an example among many of a very slow reaction called the weak interaction. This interaction is totally different from the nuclear interaction.

Electrons and neutrinos obviously do not enjoy the nuclear interaction. There is a factor of about 10^{14} by which electron- and neutrino-related reactions are slower than those involving only ordinary hadrons, under identical energy conditions. Electrons and neutrinos belong to a class of particles called *leptons*.

Weak Interactions Between Hadrons

We can (so far) identify two classes of particle: the hadrons that enjoy the strong nuclear interaction among themselves and the leptons that enjoy the weak interaction, among themselves and with hadrons, but do not enjoy the nuclear interaction at all. But the hadrons also enjoy the weak interaction among themselves even when the nuclear interaction is active. How can we possibly know this when the weak interaction is so very much weaker than the nuclear interaction? The answer is that the infinitesimal weak interaction carries a crucial distinguishing signature that enables us to recognize its effect even in the presence of the overwhelmingly stronger nuclear interaction.

Parity

This distinguishing signature takes us back to considerations of symmetry. Consider the simplest symmetry, with which we started, namely mirror symmetry which we generalize into a physical principle. This principle states that if a physical system can exist then so can its mirror image. For this to be so the system must be invariant under the symmetry operation *parity* (P). Parity is respected to a very high degree (better than 1 part in 10^{6}) by the nuclear interaction but is totally violated by the weak interaction. For example, in $n \rightarrow p + e^{-} + \bar{\nu}$ the $\bar{\nu}$ always emerges with its intrinsic spin of ½ aligned along its direction of motion: $\bar{\nu}$ is like a right-handed corkscrew. The mirror image of $n \rightarrow p + e^{-} + \bar{\nu}$ would turn the right-handed corkscrew $\bar{\nu}$ into a left-handed corkscrew $\bar{\nu}$ because mirrors turn right hands into left hands. But a left-handed $\bar{\nu}$ does not exist so neither can the mirror image of $n \rightarrow p + e^{-} + \bar{\nu}$: P is violated. This is an example of a law of nature, P-symmetry or conservation of parity, that is sometimes obeyed (by the nuclear interaction) and sometimes not (by the weak interaction). We thus see that the nuclear interaction is ambidextrous, does not carry a handedness, while the weak interaction has an intrinsic handedness. So although the nuclear interaction completely dominates, say, the collision between two protons, the tiny contribution of the weak interaction shows up through a tiny residual dependence on the handedness. The charge conjugation operation C that we noted earlier is violated too by the weak interaction (although, so far as we know, it is rigorously respected by the nuclear interaction).

Time Reversal

We now add to our catalogue of symmetry operations that of time reversal T-symmetry. Complex systems such as ourselves seem to have a very well-defined arrow of time. But the elementary laws of physics have no regard for the direction of the passage of time. (You cannot, for example, tell whether a film of the collision of two billiard balls is being run forward or back-to-front, which is T-symmetry, any more than you can tell if it was filmed directly or reflected in a mirror, which is P-symmetry.) Now although we cannot experimentally reverse the passage of time we can do something equivalent and reverse the sense of a physical process. Thus we can measure the rate of a nuclear reaction $a + b \rightarrow c + d$, bombarding particle b with particle a to give particles c and d, and then measure the rate of $c + d \rightarrow a + b$. If T-symmetry is obeyed we shall find that the intrinsic rates of the two reactions are the same. Such tests have been carried out with great care in the case of the nuclear interaction where indeed T-symmetry is respected to a very high degree. Nowhere in physics has T-violation, as such, been demonstrated although I must carefully emphasize "as such."

CPT-Symmetry

We now address the "as such." We have noted earlier the three great symmetry operations, C, P, and T, and have to some degree discussed them separately. We have also noted that the nuclear interaction respects them all individually while the weak interaction violates P and C individually. At the very heart of our understanding of the physical world is our belief that whatever happens to C, P, and T separately or in pairs the *overall symmetry CPT must be respected* – i.e., if we change all particles into anti-particles, reflect them in a mirror and run time backwards, a possible physical system yields another possible physical system. CPT-symmetry is therefore a kind of tenet of faith, a master symmetry the violation of which would be a shock from which we have no present idea how we might recover.

Particle and Anti-Particle Decay

It is this full CPT-symmetry that requires that the masses and the lifetimes of particle and anti-particle should be exactly the same. But CPT makes no such demand of precise equality of the particles' various modes of decay. Thus if X can decay in several alternative ways, into several alternative channels (say, $a + b, c + d, e + f$) its anti-particle $\overline{X}$ can certainly decay into the anti-channels $(\overline{a} + \overline{b}, \overline{c} + \overline{d}, \overline{e} + \overline{f})$. CPT-symmetry now requires that the total decay probability per unit time of X into all possible channels, that is to say the lifetime of X must be exactly the same as the lifetime of $\overline{X}$ but it does *not* demand that there should be exact equality of probability channel by channel so that the probability that X decays into channel $a + b$ could be different from the probability that $\overline{X}$ decays into $\overline{a} + \overline{b}$. This will be crucially important for us shortly. Such an inequality would, however, demand a violation of CP; if CPT-symmetry is to be respected, a violation of CP should be combined with a compensatory violation of T to preserve CPT-symmetry.

Electromagnetism and Gravitation

We have so far dwelt upon two very different interactions, the nuclear and the weak. More familiar to us in our everyday experience are the electromagnetic and gravitational interactions.

The electromagnetic interaction acts between electric charges and electric currents. It simply adds onto the nuclear and weak interactions if the particles concerned, hadrons or leptons, are electrically charged or possess magnetic moments. We do not know of any particles that enjoy only the electromagnetic interaction. At short distances (of the order of 10^{-13} cm) the electric interaction between two charged hadrons is considerably weaker than the nuclear interaction. But the nuclear interaction falls off much more rapidly with distance than the electric interaction. In fact the two become equal at separations of 6 or 7×10^{-13} cm beyond which the electric interaction rapidly dominates. The reason for this curious difference in behaviour (and the reason why the weak interaction drops off with distance even more rapidly than the nuclear), will be examined shortly.

Gravitation is the interaction with which we are most immediately familiar although it is almost unimaginably weaker than the electric interaction. If a hydrogen atom were held together by the gravitational interaction between its electron and proton, rather than by the electric interaction, it would be bigger than the universe. The reason why the gravitational interaction is so dominant on the large scale is that it operates between *all* masses, irrespective of their other properties. At large distances, the nuclear and weak interactions having become negligibly small because of their rapid fall-off with distance, gravity is all that is left to operate between electrically-neutral bodies such as the sun, planets, and spaceships. (Like the electric force, the gravitational force falls off only as the inverse square of the separation.) So far as we know, the electromagnetic and gravitational interactions respect C, P, T, and all other physical symmetries of relevance to them.

The Four Interactions and Their Mediations

In terms of our own experiences and the gleanings of our experiments, there seem to be four interactions in the particle world – nuclear, weak, electromagnetic, and gravitational. But how do these forces come about? What generates them?

We do not like the idea of action at a distance, the idea of forces that operate between two separate bodies without the mediation of an agency through which the particles communicate the news of their existence to one another. Our modern concept of force is intimately coupled to the idea of its propagation by some messenger particle that travels to and fro between the particles in question thereby, in effect, constituting their interaction. Thus everything that we know about the electromagnetic interaction is explained by the notion that the electric charges are exchanging photons, massless electrically-neutral particles of light. The quantum elaboration of this notion as Quantum Electro-Dynamics (QED) is a prodigious quantitative triumph that, for example, permits us to calculate the ratio of the magnetic moment to the angular momentum of the electron correctly to 1 part in 10^{11}.

The nuclear interaction is more complicated. It falls off rapidly at large distances, much faster than the electric and gravitational interactions. The reason for that has to do with the nature of the messenger particle of the nuclear force. Those messenger particles are the mesons of which the lightest is the pion, π, of mass about 270 times that of the electron. We believe that a neutron can interact with a proton by throwing out a π^-, thereby converting itself into a proton, the π^- later being caught by the proton which thereby becomes a neutron. So to begin with we might have a neutron to our left and a proton to our right; then a proton to our left and also a proton to our right but with a π^- "in the air" between them; then finally just a proton to our left and a neutron to our right, the communication having been effected.

But how can this be, since the pion has mass and therefore energy $E = mc^2$. There is no source for this energy. When the π^- is "in the air" where has mc^2 come from? The answer is that it has not had to come from anywhere because it has not gone anywhere: neither the initial nor the final situation contains a pion: np has simply become pn. However, for the brief intermediate state of

$p\pi^- p$ the energy $E = mc^2$ has had to be *borrowed*. This is possible by grace of a fundamental property of quantum mechanics that says we cannot simultaneously determine the values of some pairs of related quantities to better than a certain mutual precision. Energy and time are related through the equation (one of the so-called Heisenberg Uncertainty Relations) $\Delta E \Delta t = \hbar$ where $\hbar$ is the famous Planck's constant of quantum mechanics. This tells us that if we consider a system for only a finite time Δt, we cannot know its energy content – and therefore cannot tell if ΔE has been borrowed for that time Δt.

So we can borrow the energy, $E = mc^2$, to pay for our messenger pion for a time, $t = \hbar/mc^2$, during which, with the velocity of light, it could make a trip of length $c\Delta t = \hbar/m_\pi c = 1.4 \times 10^{-13}$ cm. But 1.4×10^{-13} cm is just about the distance beyond which the nuclear interaction begins to fall off rapidly, in wonderful accord with our expectation. The other heavier mesons will obviously, for the same reason, make their contributions to the nuclear interaction only at shorter distances.

The Intra-Hadronic Interaction

But what about the *intra*-hadronic forces that operate *within*, say, a proton holding its constituent quarks together? These forces we believe to be due to massless messenger particles without electrical charge, called *gluons*. These are somewhat analogous to the massless uncharged photons of the electromagnetic interaction QED. However, because these gluons, in their own mutual interactions, experience the *same* forces that they *transmit* between the quarks, the working-out of the quark-quark interaction, mediated by the exchange of massless gluons – the theory of Quantum Chromo-Dynamics (QCD) – is very different in detail from QED although it has a closely analogous starting point.

This *intra*-hadronic quark-quark interaction is called the strong interaction. It is obviously related to the *inter*-hadronic interaction that we have called the nuclear interaction. We have seen that when 3-quark neutrons and protons are sufficiently well separated they interact with each other by the exchange of pions (which are themselves a quark plus an anti-quark). The neutrons and protons under these circumstances do not directly exert single-gluon-mediated forces on each other (and cannot do so because gluons are coloured so that the exchange of a single gluon would leave the dispatching particle coloured and would colour the other; but hadrons must be colourless, as we have seen, so this process is forbidden). We see that the nuclear interaction between well-separated neutrons and protons is not the strong interaction itself, is not directly the gluon-mediated force, but is a kind of external echo of this force that itself is found *within* the intervailing neutrons and protons (and *within* the pions that mediate the nuclear interaction).

The Mediation of Gravity

We have seen that gravity is an extremely feeble interaction which makes it very difficult to study in a microscopic way. No gravitational radiation – the emission of energy by rotating or pulsating masses – has so far been detected although its emission, in good accord with the expectations of Einstein's theory of general relativity, has been rather convincingly inferred from the behaviour of a massive close binary star system. Gravity is also unique among the forces of nature in that we do not know how to describe it in the language of quantum mechanics. However, if a description of gravity is correct along the lines of Einstein's general relativity, then the mediating particle of gravity, called the *graviton*, must be massless. But more than that we do not know.

The Mediation of the Weak Interaction

We have seen that the electromagnetic interaction is mediated by (massless) photon exchange, the strong interaction by (massless) gluon exchange and its external nuclear manifestation by (massive) meson exchange while we believe that gravity is mediated by the as-yet-undiscovered (massless) graviton. What of the weak interaction?

The weak interaction is of very short range and to all intents and purposes in ordinary laboratory experiments behaves like an interaction at a point. This means that the mediating particle, called the w, positively or negatively charged, must be very massive. The phenomenology of the interaction of energetic neutrinos with matter suggests that the mass of the w must be distinctly greater than 10 times that of the proton. In fact, if we compare the electric and weak interactions they are in some ways very similar: a hadron (say a proton) interacts electrically with a charged lepton (say an electron) by emitting a photon, which is caught by the lepton. In the weak interaction the proton emits a w^+ particle, which is caught by the electron. In the electric interaction the interacting proton and electron are unchanged in the process; in the weak interaction the emission of the w^+ by the proton changes it into a neutron while the catching of the w^+ by the electron changes it into a neutrino. However, we might wonder at this stage if there might not also exist a neutral w^0 to partner the w^+ (and w^-) so that a weak proton-electron, hadron-lepton interaction could come about without an exchange of charge; indeed, such "neutral current" events are found in matter under neutrino bombardment. They, even more than the weak interactions involving exchange of charge, look similar to the photon-mediated electromagnetic interaction.

We are therefore led to ask whether there might not be a fundamental relationship between the weak and the electromagnetic interactions. They look very different: the weak interaction is very much weaker than the electromagnetic and is of such short range. But could these two differences be connected? If the intrinsic strengths of the two interactions were comparable, if the propensity of a proton to emit a W were about the same as to emit a photon, the weak interaction would still *look* much the weaker in practice. This is because the W is so heavy that by grace of the Uncertainty Principle, it can reach out only over very short distances. The particles between which it mediates have to be very close together, whereas the massless photon can reach out over large distances and pick up remote partners for the interactions that it mediates. There may be a symmetry between the weak and electromagnetic interactions but, if so, it is severely broken.

The Electroweak Interaction

At this point we should revert to our discussion about symmetry in the worlds of art and animate nature and our illustrations of ways in which symmetry is broken. Could it be that under certain circumstances (not those of our own ordinary experience) the weak and electromagnetic interactions are united but that under ordinary conditions that symmetry is badly broken?

It is easy, following our hypothesis of possible fundamental unification of the two interactions, to see under what conditions the two interactions could manifest their full relationship and the broken symmetry be restored. If the world were not our present low temperature world, a low energy world, but one in which the temperature was so high that all energies were greater than the mass of the W-particles, then the mass difference between the photon and the W would not matter. There would then be only a single interaction, the electroweak interaction, that indifferently might or might not exchange the charges of the interacting particles. Because we live in a world whose temperature is much less than the mass of the W the two low energy manifestations of the electroweak interaction *look* very different but they really belong together.

The massive neutral companion of the W^+ and W^-, which I have referred to as the W^0 and mediates the neutral weak current interactions, can be thought of as mixing with the photon. This mixing displaces it slightly from its charged companions in mass; it is renamed the Z^0; and it can join in with whatever the photon does and brings a bit of weak interaction into all electromagnetic processes.

The quantification of the electroweak unification leads to predictions of the masses of the W^+ and of the Z^0 that depend on only a single parameter. This parameter is not itself specified by the theory. It is the so-called Weinberg Angle.

The parameter can be experimentally determined in a host of different ways with completely concordant results. The clinching experiments confirming the theory were carried out at CERN, Geneva, in 1983 and provided direct physical evidence for the W and Z^0 particles.

Parity Non-Conservation in Electromagnetic Processes

An immediate consequence of the electroweak unification is that there is no such thing as a *purely electromagnetic interaction*. Anything that a photon can do, such as mediating the electric interaction, so can a Z^0. In addition, however, the Z^0 can do what photons cannot do, namely make as significant a contribution to atomic structure by mediating the interactions between electrons and the neutrons in the nucleus as it does for the electron-proton interaction. So Z^0 exchanges within atoms must play a tiny role, tiny because of the very short range of the Z^0 exchange, in determining atomic structure. But that tiny role can manifest itself because the W- and Z-mediated processes, the "weak" side of electroweak, have the intrinsic handedness that was discussed before. Thus the tiny influence of the Z^0 on atomic structure has itself a consequent handedness which will be passed on to a very slight circular polarization of the light emitted from unpolarized atoms. Such circular polarizations, and equivalent parity-non-conserving effects, have been measured in many atomic systems with results in complete accord with the expectations of the electroweak unification.

It would obviously be more satisfactory if we could understand the real origin of the Weinberg Angle, that is to say if we could understand the relationship between the electroweak unification and the rest of relevant physics. This one could do if we could find some larger unification that would link the electroweak interaction with the strong interaction in an intimate way. Such so-called Grand Unification of the strong and electroweak interactions could take place in many ways following the dictates of many alternative grand symmetry schemes. Such schemes necessarily entrain the existence of a number of particles generically called X in much the same way as the electroweak unification entrained the existence of the W and Z particles. Grand unification schemes enable us to predict the value of the Weinberg Angle as we measure it in our low temperature, low energy, world; it is a function of the mass, m_X, of the X-particle(s) and we can find good agreement with experiment for $m_X = 10^{15}$ GeV. This corresponds to a gigantic mass far beyond our ability to create in the laboratory. Using today's state-of-the-art accelerator technologies, it would require an accelerator tens of times the dimensions of the solar system.

We therefore believe that above a given temperature we may enter a grand unified world in which distinctions between quarks and leptons vanish so that they freely interconvert through the agency of the x-particles which would explain the astonishing equality of the electrical charges of electron and proton that we noticed earlier. Below that temperature the grand symmetry breaks into the strong and electroweak symmetries which, as we lower the temperature, is the situation until 10^2 GeV or so. Then, as we have already seen, the electroweak symmetry breaks, the weak and electromagnetic symmetries split apart, and we enter our familiar low-temperature world.

Proton Decay

Now although we cannot hope to make an x-particle in the laboratory and so study its supposedly even-handed mediation between quarks and leptons, we must anticipate that it can gain an extremely fleeting existence within, say, a proton. It would do that by virtue of borrowing its extraordinary mass for a very short time via the Uncertainty Principle. But if an x-particle comes into such rare and fleeting being within a proton by, say, the fusion of two of the three quarks, it may decay almost immediately, not necessarily into two more quarks but perhaps into an electron and an anti-quark. The anti-quark could then go off with the remaining original quarks as a pion: *the proton would then have decayed*.

Grand unification therefore seems to entrain the necessity of such decay of the proton itself. Quantitatively, according to the various possible symmetry schemes of grand unification, we expect the lifetime against such decay to be at least 10^{30} years. Such decay is being energetically sought in several major experiments around the world. We already know that the lifetime is longer than 10^{32} years. This argues against the simplest grand unification symmetry which has difficulty in accommodating so long a lifetime.

The Big Bang

Why are we so interested in grand unification? We have just imagined a world in which the temperature starts by being very high, and have traced the successive breaking of the symmetries as the temperature falls. But this is exactly what we think happened in the earliest instants of the universe following its coming into being in a Big Bang, the temperature falling as the universe expanded. General relativity theory specifies, in an essentially unambiguous way, how the temperature must have fallen with time after the Big Bang. After about 10^{-45} seconds it was about 10^{19} GeV (the significance of which we will deal with shortly); after about 10^{-35} seconds it was about the 10^{15} GeV below which grand unification is broken and the strong interaction distinguishes itself from the electroweak; after about 10^{-10} seconds, it was 10^2 GeV below which the electroweak symmetry is broken into the distinct electromagnetic and weak components and we enter the epoch of our own apparently unrelated forces.[3]

The Big Bang hypothesis is completely consistent with today's observed continuing expansion of the universe as a whole: the Hubble recession of the distant galaxies. It is also consistent with the microwave background radiation that fills the universe and bathes us with remarkable uniformity from all directions. It is of just the nature that we should expect from a body glowing at a temperature of about 3° K, the temperature to which the universe as a whole has now cooled. This microwave background is a kind of fossil relic of the radiation released in the Big Bang itself and the following instants. The Big Bang also very satisfactorily and quantitatively accounts for the observed cosmic abundances of the lightest elements: hydrogen, helium, and lithium.

The Big Bang theory faces two major problems: the universe is made of matter with no significant sign of anti-matter; the amount of radiation in the universe in the form of photons (chiefly in the microwave background) enormously exceeds the amount of matter (neutrons and protons) by the huge factor of 10^9. Why is this worrying? We would have expected the act of creation in the Big Bang to have been even-handed as between the production of particles and anti-particles. (How could a Creator respecting CPT-symmetry have told the difference?) We should also have expected that the Big Bang and its immediate aftermath would not have discriminated very strongly between different sorts of particle, certainly not by a factor of 10^9. How can grand unification help? In the timetable following the Big Bang, the first recognizable particles to emerge would have been the heaviest, the X-particles during the grand unification epoch prior to 10^{-35} seconds. After 10^{-35} seconds, with the breaking of the grand unification symmetry, all the X-particles would have decayed into quarks and leptons (and photons) which henceforth remained distinguishable from each other.

We follow the conviction that the act of creation should not itself have preferred particle over anti-particle. X and $\overline{X}$ were present in equal numbers prior to 10^{-35} seconds. But what of their decay? If we refer to our discussion of the decay of particle and anti-particle where, using the notation X and $\overline{X}$, I emphasized that the decay of $\overline{X}$ into channel $\overline{a} + \overline{b}$ need not have exactly the same probability as the decay of X into channel $a + b$. Now if $a + b$ is a quark-generating channel, $\overline{a} + \overline{b}$ will be an anti-quark-generating channel and so we see that there is no reason why after 10^{-35} seconds, when all the X and $\overline{X}$ have decayed, quarks and anti-quarks should be present in exactly equal numbers even though X and $\overline{X}$ were themselves exactly equally abundant.

Let us suppose that, in fact, after 10^{-35} seconds, quarks are more numerous than anti-quarks by 1 part in 10^9. The quarks and anti-quarks now set about annihilating each other until only the 1 part in 10^9 of quarks is left over. All the matter is then in the form of quarks and all the neutrons and protons into which they subsequently form themselves (this takes a few tens of microseconds) will be neutrons and protons with no anti-neutrons and no anti-protons. But the 10^9 annihilations of this cosmic Armageddon are remembered through the radiation that each has produced as the release of the mass-energy of the quarks and anti-quarks. This is the factor of 10^9 between radiation and nuclear particles that we see today and that we were seeking to explain.

So grand unification symmetry, as well as being satisfying aesthetically in bringing the three forces, strong, weak, and electromagnetic, of today's nature into a common comprehension (leaving only gravitation outside), may also simultaneously resolve the two great puzzles of the Big Bang.

Unification with Gravity?

And what of gravity, the fourth of the forces of today's world? In pursuit of unification might we not look to identify a higher symmetry that would draw also gravity into some even grander unification at some even higher temperature at some even earlier time? We at least know where to look: that 10^{-43} seconds, 10^{19} GeV in temperature, that we mentioned in the cooling curve of the universe. It corresponds after the Big Bang to the so-called Planck time within which gravity must have been the dominant interaction even though it is so relatively feeble today. Within the Planck time physics as we know it has broken down; we shall not know how to penetrate this veil and achieve the unification of gravity with the grand unified strong/electroweak interaction until we can talk about gravity in the language of quantum mechanics. This we do not yet know how to do, at least not in any unique way.

Supersymmetry, Supergravity, Superstrings

Many vigorous attempts are being made to extend the range of symmetry considerations, some of them explicitly including gravity. One such attempt concerns the enormous range of masses that we have encountered for the particles of the natural world ranging from zero for the photon and possibly for the neutrinos through the ½ MeV or so for the electron, the 1 GeV for the proton, the 100 GeV for the W and Z up to the 10^{15} GeV for the hypothetical X and the 10^{19} GeV of the Planck mass which is the mass of the observable universe at the Planck time.

It turns out to be very difficult to keep these masses apart, particularly if we go in for grand unification in which all particles belong to each other. It is very difficult to see why the mass of the W, for example, is not as great as that of the X or even the Planck mass. A way out of this problem is offered by so-called supersymmetry schemes in which every now-known particle of spin ½ is associated with a now-unknown partner of the same mass but of spin 0 while every now-known particle of spin 1 is similarly associated with a now-unknown equal-mass partner of spin ½. This supersymmetry, if the notion is correct at all, is obviously a broken symmetry because the now-unknown supersymmetric particles do not have the same masses as the known particles that they are supposed to partner. However, if the supersymmetric partners are to do their job of stabilizing the masses of the known particles they cannot themselves be very far away in mass and we might expect them to be showing up soon.

There are also so-called supergravities, with a range of possible symmetries, that contain broken supersymmetry. Very fashionable at the moment are superstring theories, in which particles are not points but lengths or loops of "string" of dimensions comparable with the Planck length of 10^{-33} cm, again involving a range of possible symmetries. Superstrings include gravity in quantum fashion and are considered by some to be the basis for a possible "theory of everything." We shall have to wait and see about that.

Arbitrariness in Symmetry-Breaking

In our illustrations of symmetry-breaking in art and in animate nature I stressed that such breaking was arbitrary in its sense – the *clocher neuf* and the *clocher vieux* could just as well have been the other way round. But so far in our discussion of symmetry-breaking in the physical domain of nature we have not had occasion to mention this point. Such arbitrariness of sense is always underlying the broken symmetries as we, *de facto*, find them. We simply use the sense of breaking as it is given to us just as we cannot now move Nefertiti's good eye.

There are, however, possibilities, particularly stemming from the very early universe, where the arbitrariness of the sense of the symmetry-breaking might have tremendous consequences visible today. This arises because the symmetry, whatever it was, might have broken in different senses in different regions of the evolving universe: in some regions Nefertiti's good eye might have been on the right and in others on the left. So long as we are in one region or the other this does not matter; she has one good eye and one bad; whichever, the effect is the same. But what when these regions meet? What happens at the boundary? One clearly cannot have a symmetry broken in two different senses at once.

Nefertiti's eye, translated into the physical language of the symmetries that may have dominated the early universe and whose broken forms we know today, leads to a range of possible consequences associated with the clash of senses of symmetry-breaking occurring at the interfaces between the regions of those differing senses. We might, for example, anticipate the very abundant generation of magnetic monopoles: classically speaking, isolated north or south poles, which have never been seen. We can also anticipate the generation of linear dislocations called cosmic strings that might have seeded the remarkable gigantic chains of galaxies and galactic clusters of which we are now becoming aware on that very different scale of the organization of the cosmos.

I am very conscious that I have not once referred to Newton. This is not so much because Newton and the *Principia* are irrelevant for the physics of today as that they are all-pervasive: the *Principia* remains the bedrock upon which modern physical science is built. Of course, much of what I have been presenting could not have been known to Newton but some of the considerations were already in his mind. Thus he clearly envisaged the existence of the very strong cohesive forces that we now understand as operating at the atomic and subatomic levels and he was clearly concerned about the mechanisms of the propagation of interactions.

NOTES

1 10^x = 1 followed by x noughts; 10^{-x} = 1 divided by 10^x: $10^4 = 10{,}000$; $10^{-4} = 1/10{,}000$.
2 He was, incidentally, reputedly the lover of not one but two queens of France, Joan of Navarre and Margaret of Burgundy. Heads of universities have come down in the world a notch or two since those days.
3 When we measure temperature T in units of energy E we are using the equivalence $E = kT$ where k is the Boltzmann's constant of thermodynamics so that T is the temperature at which particle energies E are usual by way of ordinary thermal agitation. Numerically 1 GeV is equivalent to about 10^{13} degrees K.

Contributors

A.P. FRENCH, professor of physics, Massachusetts Institute of Technology, is the author of *Introduction to Quantum Physics* (1978). He is the editor of *Einstein: A Centenary Volume* (1979) and co-editor of *Niels Bohr: A Centenary Volume* (1985).

WERNER ISRAEL is professor of physics, University of Alberta. He is a fellow of the Royal Society of Canada and received the Medal for Achievement in Physics of the Canadian Association of Physicists in 1981.

W.H. NEWTON-SMITH, Fellow of Balliol College, Oxford, is a Canadian and one of the leading philosophers of science in Britain. He is the author of three books: *The Nature of Time*, *The Rationality of Science*, and *Logic*.

D.D. RAPHAEL is emeritus professor of philosophy, University of London. He is joint editor of *Adam Smith's Lectures on Jurisprudence* (1978) and *Adam Smith's Essays on Philosophical Subjects* (1981), and the author of *Adam Smith* (1985).

STEPHEN SMALE is the 1966 Fields Medallist and recipient of the Veblen Prize. Since 1964 he has been professor of mathematics at the University of California at Berkeley.

STEVEN WEINBERG is Josey Professor of Science at the University of Texas, Austin, and winner of the Nobel Prize in Physics, 1979.

RICHARD S. WESTFALL, professor of the history of science, University of Indiana, Bloomington, is one of the leading historians of seventeenth-century science. He is the author of four books on science and religion in England. His most recent work is *Never at Rest: A Biography of Isaac Newton* (1980).

DENYS WILKINSON has recently retired as chancellor of the University of Sussex. He is the author of *Mesons in Nuclei* (1979) and *Isospin in Nuclear Physics* (1969).

Organizing Committee

L.L. CAMPBELL

B. CASTEL

K. LAKE

M. LOVE

E. MANOUKIAN

W. MCLATCHIE, CHAIR

D. NORMAN

M.J. STOTT